Cram101

Cram101.com for Practice Tests

Textbook Outlines, Highlights, and Practice Quizzes

Anatomy and Physiology for Midwives

by Jane Coad, Melvyn Dunstall, 3rd Edition

All "Just the Facts101" Material Written or Prepared by Cram101 Publishing

Title Page

WHY STOP HERE... THERE'S MORE ONLINE

With technology and experience, we've developed tools that make studying easier and efficient. Like this Cram101 textbook notebook, **Cram101.com** offers you the highlights from every chapter of your actual textbook. However, unlike this notebook, **Cram101.com** gives you practice tests for each of the chapters. You also get access to in-depth reference material for writing essays and papers.

By purchasing this book, you get 50% off the normal subscription free!. Just enter the promotional code **'DK73DW20612'** on the Cram101.com registration screen.

CRAM101.COM FEATURES:

Outlines & Highlights
Just like the ones in this notebook, but with links to additional information.

Integrated Note Taking
Add your class notes to the Cram101 notes, print them and maximize your study time.

Problem Solving
Step-by-step walk throughs for math, stats and other disciplines.

Practice Exams
Five different test taking formats for every chapter.

Easy Access
Study any of your books, on any computer, anywhere.

Unlimited Textbooks
All the features above for virtually all your textbooks, just add them to your account at no additional cost.

Be sure to use the promo code above when registering on Cram101.com to get 50% off your membership fees.

STUDYING MADE EASY

This Cram101 notebook is designed to make studying easier and increase your comprehension of the textbook material. Instead of starting with a blank notebook and trying to write down everything discussed in class lectures, you can use this Cram101 textbook notebook and annotate your notes along with the lecture.

Our goal is to give you the best tools for success.

For a supreme understanding of the course, pair your notebook with our online tools. Should you decide you prefer Cram101.com as your study tool,

we'd like to offer you a trade...

Our Trade In program is a simple way for us to keep our promise and provide you the best studying tools, regardless of where you purchased your Cram101 textbook notebook. As long as your notebook is in *Like New Condition**, you can send it back to us and we will immediately give you a Cram101.com account free for 120 days!

Let The *Trade In* Begin!

THREE SIMPLE STEPS TO TRADE:

1. Go to www.cram101.com/tradein and fill out the packing slip information.

2. Submit and print the packing slip and mail it in with your Cram101 textbook notebook.

3. Activate your account after you receive your email confirmation.

---

* Books must be returned in *Like New Condition*, meaning there is no damage to the book including, but not limited to; ripped or torn pages, markings or writing on pages, or folded / creased pages. Upon receiving the book, Cram101 will inspect it and reserves the right to terminate your free Cram101.com account and return your textbook notebook at the owners expense.

"Just the Facts101" is a Cram101 publication and tool designed to give you all the facts from your textbooks. Visit Cram101.com for the full practice test for each of your chapters for virtually any of your textbooks.

Cram101 has built custom study tools specific to your textbook. We provide all of the factual testable information and unlike traditional study guides, we will never send you back to your textbook for more information.

YOU WILL NEVER HAVE TO HIGHLIGHT A BOOK AGAIN!

Cram101 StudyGuides
All of the information in this StudyGuide is written specifically for your textbook. We include the key terms, places, people, and concepts... the information you can expect on your next exam!

Want to take a practice test?
Throughout each chapter of this StudyGuide you will find links to cram101.com where you can select specific chapters to take a complete test on, or you can subscribe and get practice tests for up to 12 of your textbooks, along with other exclusive cram101.com tools like problem solving labs and reference libraries.

Cram101.com
Only cram101.com gives you the outlines, highlights, and PRACTICE TESTS specific to your textbook. Cram101.com is an online application where you'll discover study tools designed to make the most of your limited study time.

By purchasing this book, you get 50% off the normal subscription free!. Just enter the promotional code **'DK73DW20612'** on the Cram101.com registration screen.

www.Cram101.com

Copyright © 2012 by Cram101, Inc. All rights reserved.
"Just the FACTS101"®, "Cram101"® and "Never Highlight a Book Again!"® are registered trademarks of Cram101, Inc.
ISBN(s): 9781478419495. PUBX-1.20121214

Learning System

Anatomy and Physiology for Midwives
Jane Coad, Melvyn Dunstall, 3rd

CONTENTS

1. Introduction to physiology 5
2. The reproductive and urinary systems 21
3. Endocrinology 37
4. Reproductive cycles 55
5. Sexual differentiation and behaviour 69
6. Fertilization 75
7. Overview of human genetics and genetic disorders 87
8. The placenta 97
9. Embryo development and fetal growth 107
10. Overview of immunology 119
11. Physiological adaptation to pregnancy 131
12. Maternal nutrition and health 146
13. Physiology of parturition 157
14. The puerperium 171
15. The transition to neonatal life 179
16. Lactation and infant nutrition 189

Chapter 1. Introduction to physiology
CHAPTER OUTLINE: KEY TERMS, PEOPLE, PLACES, CONCEPTS

_____ Cell

_____ Ejaculatory duct

_____ Cardiac muscle

_____ Smooth muscle

_____ Adipose tissue

_____ Connective tissue

_____ Vasoconstriction

_____ Neuron

_____ Ectopic pregnancy

_____ Glial cell

_____ Homeostasis

_____ Motility

_____ Nervous system

_____ Thermoregulation

_____ Autonomic nervous system

_____ Neurotransmitter

_____ Anabolism

_____ Endocrine system

_____ Hypothalamus

Chapter 1. Introduction to physiology
CHAPTER OUTLINE: KEY TERMS, PEOPLE, PLACES, CONCEPTS

_____	Small intestine
_____	Gut-associated lymphoid tissue
_____	Immune system
_____	Large intestine
_____	Interaction
_____	Asphyxia
_____	Respiratory system
_____	Oxygen
_____	Blood vessel
_____	Heart development
_____	Pulmonary circulation
_____	Systemic circulation
_____	Vasopressin
_____	Baroreceptor
_____	Blood flow
_____	Blood pressure
_____	Blood-brain barrier
_____	Cardiac output
_____	Chemoreceptor

Chapter 1. Introduction to physiology

CHAPTER OUTLINE: KEY TERMS, PEOPLE, PLACES, CONCEPTS

- Coagulation
- Cord blood
- Coronary circulation
- Erythropoietin
- Platelet
- Sinoatrial node
- Blood cell
- Blood test
- Lymph node
- Lymphatic system
- Lymphocyte
- Metabolism
- Carbohydrate
- Glycogen
- Glycolysis
- Fatty acid
- Glycogenolysis
- Ketone bodies
- Protein metabolism

Chapter 1. Introduction to physiology

CHAPTER OUTLINE: KEY TERMS, PEOPLE, PLACES, CONCEPTS

- Absorptive state
- Blood sugar
- Blood sugar regulation

CHAPTER HIGHLIGHTS & NOTES: KEY TERMS, PEOPLE, PLACES, CONCEPTS

Cell	The cell is the functional basic unit of life. It was discovered by Robert Hooke and is the functional unit of all known living organisms. It is the smallest unit of life that is classified as a living thing, and is often called the building block of life.
Ejaculatory duct	The ejaculatory ducts (ductus ejaculatorii) are paired structures in male anatomy. Each ejaculatory duct is formed by the union of the vas deferens with the duct of the seminal vesicle. They pass through the prostate, and open into the urethra at the Colliculus seminalis.
Cardiac muscle	Cardiac muscle is a type of involuntary striated muscle found in the walls and histological foundation of the heart, specifically the myocardium. Cardiac muscle is one of three major types of muscle, the others being skeletal and smooth muscle. The cells that comprise cardiac muscle, called cardiomyocytes or myocardiocyteal muscle cells, can contain one, two, or very rarely three or four cell nuclei.
Smooth muscle	Smooth muscle is an involuntary non-striated muscle. It is divided into two sub-groups; the single-unit (unitary) and multiunit smooth muscle. Within single-unit smooth muscle tissues, the autonomic nervous system innervates a single cell within a sheet or bundle and the action potential is propagated by gap junctions to neighboring cells such that the whole bundle or sheet contracts as a syncytium (i.e., a multinucleate mass of cytoplasm that is not separated into cells).
Adipose tissue	In histology, adipose tissue is loose connective tissue composed of adipocytes. It is technically composed of roughly only 80% fat; fat in its solitary state exists in the liver and muscles. Adipose tissue is derived from lipoblasts.
Connective tissue	Connective tissue is a fibrous tissue. It is one of the four traditional classes of tissues (the others being epithelial, muscle, and nervous tissue).

Chapter 1. Introduction to physiology

CHAPTER HIGHLIGHTS & NOTES: KEY TERMS, PEOPLE, PLACES, CONCEPTS

Vasoconstriction	Vasoconstriction is the narrowing of the blood vessels resulting from contraction of the muscular wall of the vessels, particularly the large arteries and small arterioles. The process is the opposite of vasodilation, the widening of blood vessels. The process is particularly important in staunching hemorrhage and acute blood loss.
Neuron	A neuron is an electrically excitable cell that processes and transmits information by electrical and chemical signaling. Chemical signaling occurs via synapses, specialized connections with other cells. Neurons connect to each other to form networks.
Ectopic pregnancy	An ectopic pregnancy, is a complication of pregnancy in which the embryo implants outside the uterine cavity. With rare exceptions, ectopic pregnancies are not viable. Furthermore, they are dangerous for the parent, since internal haemorrhage is a life threatening complication.
Glial cell	Glial cells, sometimes called neuroglia or simply glia, are non-neuronal cells that maintain homeostasis, form myelin, and provide support and protection for the brain's neurons. In the human brain, there is roughly one glia for every neuron with a ratio of about two neurons for every three glia in the cerebral gray matter.
Homeostasis	Homeostasis is the property of a system that regulates its internal environment and tends to maintain a stable, constant condition of properties such as temperature or pH. It can be either an open or closed system. It was defined by Claude Bernard and later by Walter Bradford Cannon in 1926, 1929 and 1932. Typically used to refer to a living organism, the concept came from that of milieu intérieur that was created by Claude Bernard and published in 1865. Multiple dynamic equilibrium adjustment and regulation mechanisms make homeostasis possible.
Motility	Motility is a biological term which refers to the ability to move spontaneously and actively, consuming energy in the process. Most animals are motile but the term applies to unicellular and simple multicellular organisms, as well as to some mechanisms of fluid flow in multicellular organs, in addition to animal locomotion. Motile marine animals are commonly called free-swimming.
Nervous system	The nervous system is an organ system containing a network of specialized cells called neurons that coordinate the actions of an animal and transmit signals between different parts of its body. In most animals the nervous system consists of two parts, central and peripheral. The central nervous system of vertebrates (such as humans) contains the brain, spinal cord, and retina. The peripheral nervous system consists of sensory neurons, clusters of neurons called ganglia, and nerves connecting them to each other and to the central nervous system. These regions are all interconnected by means of complex neural pathways.

Chapter 1. Introduction to physiology

CHAPTER HIGHLIGHTS & NOTES: KEY TERMS, PEOPLE, PLACES, CONCEPTS

Thermoregulation	Thermoregulation is the ability of an organism to keep its body temperature within certain boundaries, even when the surrounding temperature is very different. This process is one aspect of homeostasis: a dynamic state of stability between an animal's internal environment and its external environment (the study of such processes in zoology has been called ecophysiology or physiological ecology). If the body is unable to maintain a normal temperature and it increases significantly above normal, a condition known as hyperthermia occurs. This occurs when the body is exposed to constant temperatures of approximately 55° C, any prolonged exposure (longer than a few hours) at this temperature and up to around 70° C death is almost inevitable. The opposite condition, when body temperature decreases below normal levels, is known as hypothermia.
Autonomic nervous system	The autonomic nervous system is the part of the peripheral nervous system that acts as a control system functioning largely below the level of consciousness, and controls visceral functions. The Autonomic nervous system affects heart rate, digestion, respiration rate, salivation, perspiration, diameter of the pupils, micturition (urination), and sexual arousal. Whereas most of its actions are involuntary, some, such as breathing, work in tandem with the conscious mind.
Neurotransmitter	Neurotransmitters are endogenous chemicals which transmit signals from a neuron to a target cell across a synapse. Neurotransmitters are packaged into synaptic vesicles clustered beneath the membrane on the presynaptic side of a synapse, and are released into the synaptic cleft, where they bind to receptors in the membrane on the postsynaptic side of the synapse. Release of neurotransmitters usually follows arrival of an action potential at the synapse, but may also follow graded electrical potentials.
Anabolism	Anabolism is the set of metabolic pathways that construct molecules from smaller units. These reactions require energy. One way of categorizing metabolic processes, whether at the cellular, organ or organism level is as 'anabolic' or as 'catabolic', which is the opposite.
Endocrine system	The endocrine system is the system of glands, each of which secretes a type of hormone directly into the bloodstream to regulate the body. The endocrine system is in contrast to the exocrine system, which secretes its chemicals using ducts. It derives from the Greek words 'endo' meaning inside, within, and 'crinis' for secrete.
Hypothalamus	The hypothalamus is a portion of the brain that contains a number of small nuclei with a variety of functions. One of the most important functions of the hypothalamus is to link the nervous system to the endocrine system via the pituitary gland (hypophysis).
	The hypothalamus is located below the thalamus, just above the brain stem.

Chapter 1. Introduction to physiology

CHAPTER HIGHLIGHTS & NOTES: KEY TERMS, PEOPLE, PLACES, CONCEPTS

Gut-associated lymphoid tissue	The digestive tract's immune system is often referred to as gut-associated lymphoid tissue and works to protect the body from invasion. GALT is an example of mucosa-associated lymphoid tissue. The digestive tract is an important component of the body's immune system.
Immune system	The immune system is a system of biological structures and processes within an organism that protects against disease. To function properly, an immune system must detect a wide variety of agents, from viruses to parasitic worms, and distinguish them from the organism's own healthy tissue. Pathogens can rapidly evolve and adapt to avoid detection and neutralization by the immune system.
Large intestine	The large intestine is the third-to-last part of the digestive system in vertebrate animals.
Interaction	Interaction is a kind of action that occurs as two or more objects have an effect upon one another. The idea of a two-way effect is essential in the concept of interaction, as opposed to a one-way causal effect. A closely related term is interconnectivity, which deals with the interactions of interactions within systems: combinations of many simple interactions can lead to surprising emergent phenomena.
Asphyxia	Asphyxia is a condition of severely deficient supply of oxygen to the body that arises from being unable to breathe normally. An example of asphyxia is choking. Asphyxia causes generalized hypoxia, which primarily affects the tissues and organs.
Respiratory system	The respiratory system is the biological system of an organism that introduces respiratory gases to the interior and performs gas exchange. In humans and other mammals, the anatomical features of the respiratory system include airways, lungs, and the respiratory muscles. Molecules of oxygen and carbon dioxide are passively exchanged, by diffusion, between the gaseous external environment and the blood.
Oxygen	Oxygen is the element with atomic number 8 and represented by the symbol O. At standard temperature and pressure, two atoms of the element bind to form dioxygen, a colorless, odorless, tasteless diatomic gas with the formula O_2. Oxygen is a member of the chalcogen group on the periodic table, and is a highly reactive nonmetallic element that readily forms compounds (notably oxides) with almost all other elements.

Chapter 1. Introduction to physiology

CHAPTER HIGHLIGHTS & NOTES: KEY TERMS, PEOPLE, PLACES, CONCEPTS

Blood vessel	The blood vessels are the part of the circulatory system that transports blood throughout the body. There are three major types of blood vessels: the arteries, which carry the blood away from the heart; the capillaries, which enable the actual exchange of water and chemicals between the blood and the tissues; and the veins, which carry blood from the capillaries back toward the heart. The arteries and veins have different structures, veins having two layers and arteries having three:•Tunica intima (the thinnest layer): a single layer of simple squamous endothelial cells glued by a polysaccharide intercellular matrix, surrounded by a thin layer of subendothelial connective tissue interlaced with a number of circularly arranged elastic bands called the internal elastic lamina.•Tunica media (the thickest layer): circularly arranged elastic fiber, connective tissue, polysaccharide substances, the second and third layer are separated by another thick elastic band called external elastic lamina.
Heart development	The heart is the first functional organ in a vertebrate embryo. There are 5 stages to heart development. The lateral plate mesoderm delaminates to form two layers: the dorsal somatic (parietal) mesoderm and the ventral splanchnic (visceral) mesoderm.
Pulmonary circulation	Pulmonary circulation is the half portion of the cardiovascular system which carries oxygen-depleted blood away from the heart, to the lungs, and returns oxygenated (oxygen-rich) blood back to the heart. The term pulmonary circulation is readily paired and contrasted with the systemic circulation. A separate system known as the bronchial circulation supplies blood to the tissue of the larger airways of the lung.
Systemic circulation	Systemic circulation is the part of the cardiovascular system which carries oxygenated blood away from the heart to the body, and returns deoxygenated blood back to the heart. This physiologic theory of circulation was first described by William Harvey. This term is opposed and contrasted to the term pulmonary circulation first proposed by Ibn al-Nafis.
Vasopressin	Arginine vasopressin also known as vasopressin, argipressin or antidiuretic hormone (ADH), is a neurohypophysial hormone found in most mammals. Vasopressin is responsible for increasing water absorption in the collecting ducts of the kidney nephron. Vasopressin increases water permeability of kidney collecting duct by inducing translocation of aquaporin-CD water channels in the kidney nephron collecting duct plasma membrane.
Baroreceptor	Baroreceptors are sensors located in the blood vessels of several mammals. They are a type of mechanoreceptor that detects the pressure of blood flowing through them, and can send messages to the central nervous system to increase or decrease total peripheral resistance and cardiac output.

Chapter 1. Introduction to physiology

CHAPTER HIGHLIGHTS & NOTES: KEY TERMS, PEOPLE, PLACES, CONCEPTS

Blood flow	Blood flow is the continuous running of blood in the cardiovascular system. The human body is made up of several processes all carrying out various functions. We have the gastrointestinal system which aids the digestion and the absorption of food.
Blood pressure	Blood pressure is the pressure exerted by circulating blood upon the walls of blood vessels, and is one of the principal vital signs. When used without further specification, 'blood pressure' usually refers to the arterial pressure of the systemic circulation. During each heartbeat, blood pressure varies between a maximum (systolic) and a minimum (diastolic) pressure.
Blood-brain barrier	The blood-brain barrier is a separation of circulating blood and the brain extracellular fluid (BECF) in the central nervous system (CNS). It occurs along all capillaries and consists of tight junctions around the capillaries that do not exist in normal circulation. Endothelial cells restrict the diffusion of microscopic objects (e.g. bacteria) and large or hydrophilic molecules into the cerebrospinal fluid (CSF), while allowing the diffusion of small hydrophobic molecules (O_2, hormones, CO_2). Cells of the barrier actively transport metabolic products such as glucose across the barrier with specific proteins. This barrier also includes a thick basement membrane and astrocytic endfeet.
Cardiac output	Cardiac output is the volume of blood being pumped by the heart, in particular by a left or right ventricle in the time interval of one minute. CO may be measured in many ways, for example dm^3/min (1 dm^3 equals 1000 cm^3 or 1 litre). Q is furthermore the combined sum of output from the right ventricle and the output from the left ventricle during the phase of systole of the heart.
Chemoreceptor	A chemoreceptor, is a sensory receptor that transduces a chemical signal into an action potential. In more general terms, a chemosensor detects certain chemical stimuli in the environment. There are two main classes of the chemosensor: direct and distance.
Coagulation	Coagulation is a complex process by which blood forms clots. It is an important part of hemostasis, the cessation of blood loss from a damaged vessel, wherein a damaged blood vessel wall is covered by a platelet and fibrin-containing clot to stop bleeding and begin repair of the damaged vessel. Disorders of coagulation can lead to an increased risk of bleeding (hemorrhage) or obstructive clotting (thrombosis).
Cord blood	Umbilical cord blood is blood that remains in the placenta and in the attached umbilical cord after childbirth. Cord blood is collected because it contains stem cells which can be used to treat hematopoietic and genetic disorders.

Chapter 1. Introduction to physiology

CHAPTER HIGHLIGHTS & NOTES: KEY TERMS, PEOPLE, PLACES, CONCEPTS

Coronary circulation	Coronary circulation is the circulation of blood in the blood vessels of the heart muscle (the myocardium). The vessels that deliver oxygen-rich blood to the myocardium are known as coronary arteries. The vessels that remove the deoxygenated blood from the heart muscle are known as cardiac veins.
Erythropoietin	Erythropoietin, ??r?θr?'p??t?n, or ??ri?θr?-) or EPO, is a glycoprotein hormone that controls erythropoiesis, or red blood cell production. It is a cytokine (protein signaling molecule) for erythrocyte (red blood cell) precursors in the bone marrow. Also called hematopoietin or hemopoietin, it is produced by interstitial fibroblasts in the kidney in close association with peritubular capillary and tubular epithelial cells.
Platelet	Platelets, or thrombocytes, are small, irregularly shaped clear cell fragments (i.e. cells that do not have a nucleus containing DNA), 2-3 μm in diameter, which are derived from fragmentation of precursor megakaryocytes. The average lifespan of a platelet is normally just 5 to 9 days. Platelets are a natural source of growth factors. They circulate in the blood of mammals and are involved in hemostasis, leading to the formation of blood clots.
Sinoatrial node	The sinoatrial node is the impulse-generating (pacemaker) tissue located in the right atrium of the heart, and thus the generator of normal sinus rhythm. It is a group of cells positioned on the wall of the right atrium, near the entrance of the superior vena cava. These cells are modified cardiac myocytes.
Blood cell	A blood cell, is a cell produced by haematopoiesis and normally found in blood. In mammals, these fall into three general categories:•Red blood cells -- Erythrocytes•White blood cells -- Leukocytes•Platelets -- Thrombocytes. Together, these three kinds of blood cells add up to a total 45% of the blood tissue by volume, with the remaining 55% of the volume composed of plasma, the liquid component of blood. This volume percentage (e.g., 45%) of cells to total volume is called hematocrit, determined by centrifuge or flow cytometry.
Blood test	A blood test is a laboratory analysis performed on a blood sample that is usually extracted from a vein in the arm using a needle, or via fingerprick. Blood tests are used to determine physiological and biochemical states, such as disease, mineral content, drug effectiveness, and organ function. They are also used in drug tests.
Lymph node	A lymph node is a small ball or an oval-shaped organ of the immune system, distributed widely throughout the body including the armpit and stomach and linked by lymphatic vessels. Lymph nodes are garrisons of B, T and other immune cells.

Chapter 1. Introduction to physiology

CHAPTER HIGHLIGHTS & NOTES: KEY TERMS, PEOPLE, PLACES, CONCEPTS

Lymphatic system	The lymphatic system is a part of the circulatory system, comprising a network of conduits called lymphatic vessels that carry a clear fluid called lymph unidirectionally towards the heart. The lymphatic system was first described independently by Olaus Rudbeck and Thomas Bartholin. The lymph system is not a closed system.
Lymphocyte	A lymphocyte is a type of white blood cell in the vertebrate immune system. Under the microscope, lymphocytes can be divided into large granular lymphocytes and small lymphocytes. Large granular lymphocytes include natural killer cells (NK cells). Small lymphocytes consist of T cells and B cells.
Metabolism	Metabolism is the set of chemical reactions that happen in the cells of living organisms to sustain life. These processes allow organisms to grow and reproduce, maintain their structures, and respond to their environments. The word metabolism can also refer to all chemical reactions that occur in living organisms, including digestion and the transport of substances into and between different cells, in which case the set of reactions within the cells is called intermediary metabolism or intermediate metabolism.
Carbohydrate	A carbohydrate is an organic compound that consists only of carbon, hydrogen, and oxygen, usually with a hydrogen:oxygen atom ratio of 2:1 (as in water); in other words, with the empirical formula $C_m(H_2O)_n$. (Some exceptions exist; for example, deoxyribose, a component of DNA, has the empirical formula $C_5H_{10}O_4$). Carbohydrates are not technically hydrates of carbon.
Glycogen	Glycogen is a multibranched polysaccharide that serves as a form of energy storage in animals and fungi. In humans, glycogen is made and stored primarily in the cells of the liver and the muscles, and functions as the secondary long-term energy storage (with the primary energy stores being fats held in adipose tissue). Glycogen can also be made via glycogenesis within the brain and stomach.
Glycolysis	Glycolysis is the metabolic pathway that converts glucose $C_6H_{12}O_6$, into pyruvate, $CH_3COCOO^- + H^+$. The free energy released in this process is used to form the high-energy compounds ATP (adenosine triphosphate) and NADH (reduced nicotinamide adenine dinucleotide). Glycolysis is a definite sequence of ten reactions involving ten intermediate compounds (one of the steps involves two intermediates).
Fatty acid	In chemistry, especially biochemistry, a fatty acid is a carboxylic acid with a long aliphatic tail (chain), which is either saturated or unsaturated. Most naturally occurring fatty acids have a chain of an even number of carbon atoms, from 4 to 28. Fatty acids are usually derived from triglycerides or phospholipids.

Chapter 1. Introduction to physiology

CHAPTER HIGHLIGHTS & NOTES: KEY TERMS, PEOPLE, PLACES, CONCEPTS

Glycogenolysis	Glycogenolysis is the break down of glycogen to glucose-1-phosphate and glucose. Glycogen branches are catabolized by the sequential removal of glucose monomers via phosphorolysis, by the enzyme glycogen phosphorylase.
	The overall reaction for the breakdown of glycogen to glucose-1-phosphate is:
	$\text{glycogen}_{(n\ \text{residues})} + P_i \longleftrightarrow \text{glycogen}_{(n-1\ \text{residues})} + \text{glucose-1-phosphate}$
	Here, glycogen phosphorylase cleaves the bond linking a terminal glucose residue to a glycogen branch by substitution of a phosphoryl group for the α[1→4] linkage.
Ketone bodies	Ketone bodies are three water-soluble compounds that are produced as by-products when fatty acids are broken down for energy in the liver. Two of the three are used as a source of energy in the heart and brain while the third is a waste product excreted from the body. In the brain, they are a vital source of energy during fasting.
Protein metabolism	Protein metabolism denotes the various biochemical processes responsible for the synthesis of proteins and amino acids, and the breakdown of proteins (and other large molecules, too) by catabolism.
	Main article: Protein biosynthesis.
	Protein biosynthesis relies on four processes;•amino acid synthesis•RNA synthesis•transcription•translation
	Protein anabolism is the process by which protein are formed from amino acids (aka anabolic amino acid synthesis).
Absorptive state	Absorptive state is the period in which the gastrointestinal tract is full and the anabolic processes exceed catabolism. The fuel used for this process is glucose.
	Carbohydrates - Simple sugars are sent to the liver where they are converted to glucose.
Blood sugar	The blood sugar concentration or blood glucose level is the amount of glucose (sugar) present in the blood of a human or animal. Normally in mammals, the body maintains the blood glucose level at a reference range between about 3.6 and 5.8 mM (mmol/L, i.e., millimoles/liter), or 64.8 and 104.4 mg/dL. The human body naturally tightly regulates blood glucose levels as a part of metabolic homeostasis.

Chapter 1. Introduction to physiology

CHAPTER HIGHLIGHTS & NOTES: KEY TERMS, PEOPLE, PLACES, CONCEPTS

Blood sugar regulation	Blood sugar regulation is the process by which the levels of blood sugar, primarily glucose, are maintained by the body.
	Blood sugar levels are regulated by negative feedback in order to keep the body in homeostasis. The levels of glucose in the blood are monitored by the cells in the pancreas's Islets of Langerhans.

CHAPTER QUIZ: KEY TERMS, PEOPLE, PLACES, CONCEPTS

1. _____ is a multibranched polysaccharide that serves as a form of energy storage in animals and fungi. In humans, _____ is made and stored primarily in the cells of the liver and the muscles, and functions as the secondary long-term energy storage (with the primary energy stores being fats held in adipose tissue). _____ can also be made via _____esis within the brain and stomach.

    a. Glycogen
    b. Gut loading
    c. Harris-Benedict equation
    d. Health effects of chocolate

2. The _____ is the biological system of an organism that introduces respiratory gases to the interior and performs gas exchange. In humans and other mammals, the anatomical features of the _____ include airways, lungs, and the respiratory muscles. Molecules of oxygen and carbon dioxide are passively exchanged, by diffusion, between the gaseous external environment and the blood.

    a. Respiratory system
    b. Autonomic dysreflexia
    c. Eclampsia
    d. Electric shock

3. _____ is a kind of action that occurs as two or more objects have an effect upon one another. The idea of a two-way effect is essential in the concept of _____, as opposed to a one-way causal effect. A closely related term is interconnectivity, which deals with the _____s of _____s within systems: combinations of many simple _____s can lead to surprising emergent phenomena.

    a. International Cannabinoid Research Society
    b. International Nonproprietary Name
    c. International Psychopharmacology Algorithm Project
    d. Interaction

Chapter 1. Introduction to physiology

CHAPTER QUIZ: KEY TERMS, PEOPLE, PLACES, CONCEPTS

4. In histology, _____ is loose connective tissue composed of adipocytes. It is technically composed of roughly only 80% fat; fat in its solitary state exists in the liver and muscles. _____ is derived from lipoblasts.

    a. Android Fat Distribution
    b. Empty sella syndrome
    c. Adipose tissue
    d. Adjustable gastric band

5. The _____ is the system of glands, each of which secretes a type of hormone directly into the bloodstream to regulate the body. The _____ is in contrast to the exocrine system, which secretes its chemicals using ducts. It derives from the Greek words 'endo' meaning inside, within, and 'crinis' for secrete.

    a. Adrenal gland
    b. Adrenal medulla
    c. Adrenalectomy
    d. Endocrine system

ANSWER KEY
Chapter 1. Introduction to physiology

1. a
2. a
3. d
4. c
5. d

You can take the complete Chapter Practice Test

for Chapter 1. Introduction to physiology
on all key terms, persons, places, and concepts.

Online 99 Cents

http://www.epub5042.32.20612.1.cram101.com/

Use www.Cram101.com for all your study needs

including Cram101's online interactive problem solving labs in

chemistry, statistics, mathematics, and more.

Chapter 2. The reproductive and urinary systems
CHAPTER OUTLINE: KEY TERMS, PEOPLE, PLACES, CONCEPTS

	Kidney
	Urinary system
	Glomerulus
	Nephron
	Fatty acid
	Aldosterone
	Distal convoluted tubule
	Parathyroid hormone
	Reabsorption
	Selective reabsorption
	Secretion
	Ureter
	Circumcision
	Female genital mutilation
	Urinary tract infection
	Urine
	Chlamydia trachomatis
	Fallopian tube
	Oviduct

Chapter 2. The reproductive and urinary systems
CHAPTER OUTLINE: KEY TERMS, PEOPLE, PLACES, CONCEPTS

_____ Uterus

_____ Ligament

_____ Endometrium

_____ Methotrexate

_____ Myometrium

_____ Salpingectomy

_____ Ectopic pregnancy

_____ Gonadotropin

_____ Cervix

_____ Pancreas

_____ Perimetrium

_____ Cancer

_____ Clitoris

_____ Hymen

_____ Labia majora

_____ Labia minora

_____ Vulva

_____ Pelvic floor

_____ Pelvic inflammatory disease

Chapter 2. The reproductive and urinary systems

CHAPTER OUTLINE: KEY TERMS, PEOPLE, PLACES, CONCEPTS

_____ Cremaster muscle

_____ Scrotum

_____ Seminiferous tubule

_____ Tunica albuginea

_____ Epididymis

_____ Vas deferens

_____ Ejaculatory duct

_____ Foreskin

_____ Gamete

_____ Gametogenesis

_____ Penis

_____ Semen

_____ Seminal vesicle

_____ Follicle-stimulating hormone

_____ Luteinizing hormone

_____ Mitosis

_____ Sertoli cell

_____ Spermatocyte

_____ Spermatogenesis

Chapter 2. The reproductive and urinary systems
CHAPTER OUTLINE: KEY TERMS, PEOPLE, PLACES, CONCEPTS

	Spermiogenesis
	Testosterone
	Hypothalamus
	Leydig cell

CHAPTER HIGHLIGHTS & NOTES: KEY TERMS, PEOPLE, PLACES, CONCEPTS

Kidney	The kidneys are organs with several functions. They are seen in many types of animals, including vertebrates and some invertebrates. They are an essential part of the urinary system and also serve homeostatic functions such as the regulation of electrolytes, maintenance of acid-base balance, and regulation of blood pressure. They serve the body as a natural filter of the blood, and remove wastes which are diverted to the urinary bladder. In producing urine, the kidneys excrete wastes such as urea and ammonium; the kidneys also are responsible for the reabsorption of water, glucose, and amino acids. The kidneys also produce hormones including calcitriol, renin, and erythropoietin.
Urinary system	The urinary system is the organ system that produces, stores, and eliminates urine. In humans it includes two kidneys, two ureters, the bladder, the urethra, and two sphincter muscles.
Glomerulus	In the kidney, a tubular structure called the nephron filters blood to form urine. At the beginning of the nephron, the glomerulus is a network (tuft) of capillaries that performs the first step of filtering blood. The glomerulus is surrounded by Bowman's capsule.
Nephron	Nephron is the basic structural and functional unit of the kidney. Its chief function is to regulate the concentration of water and soluble substances like sodium salts by filtering the blood, reabsorbing what is needed and excreting the rest as urine. A nephron eliminates wastes from the body, regulates blood volume and blood pressure, controls levels of electrolytes and metabolites, and regulates blood pH.

Chapter 2. The reproductive and urinary systems

CHAPTER HIGHLIGHTS & NOTES: KEY TERMS, PEOPLE, PLACES, CONCEPTS

Fatty acid	In chemistry, especially biochemistry, a fatty acid is a carboxylic acid with a long aliphatic tail (chain), which is either saturated or unsaturated. Most naturally occurring fatty acids have a chain of an even number of carbon atoms, from 4 to 28. Fatty acids are usually derived from triglycerides or phospholipids. When they are not attached to other molecules, they are known as 'free' fatty acids.
Aldosterone	Aldosterone is a yellow steroid hormone (mineralocorticoid family) produced by the outer-section (zona glomerulosa) of the adrenal cortex in the adrenal gland, and acts mainly on the distal tubules and collecting ducts of the nephron, the functioning unit of the kidney, to cause the conservation of sodium, secretion of potassium, increased water retention, and increased blood pressure. The overall effect of aldosterone is to increase reabsorption of ions and water in the kidney -- increasing blood volume and, therefore, increasing blood pressure. Drugs that interfere with the secretion or action of aldosterone are in use as antihypertensives.
Distal convoluted tubule	The distal convoluted tubule is a portion of kidney nephron between the loop of Henle and the collecting duct system. It is partly responsible for the regulation of potassium, sodium, calcium, and pH. It is the primary site for the kidneys' hormone based regulation of calcium (Ca). On its apical surface (lumen side), cells of the DCT have a thiazide-sensitive Na-Cl cotransporter and are permeable to Ca, via TRPV5 channel.
Parathyroid hormone	Parathyroid hormone parathormone or parathyrin, is secreted by the parathyroid glands as a polypeptide containing 84 amino acids. It acts to increase the concentration of calcium (Ca^{2+}) in the blood, whereas calcitonin (a hormone produced by the parafollicular cells (C cells) of the thyroid gland) acts to decrease calcium concentration. PTH acts to increase the concentration of calcium in the blood by acting upon parathyroid hormone receptor in three parts of the body: PTH half-life is approximately 4 minutes.
Reabsorption	In physiology, reabsorption is the flow of glomerular filtrate from the proximal tubule of the nephron into the peritubular capillaries, or from the urine into the blood. It is termed 'reabsorption' because this is technically the second time that the nutrients in question are being absorbed into the blood, the first time being from the small intestine into the villi. This happens as a result of sodium transport from the lumen into the blood by the Na+/K+ ATPase in the basolateral membrane of the epithelial cells.
Selective reabsorption	Selective reabsorption takes place in the proximal convoluted tubule (PCT) of the kidney. It is the process by which useful substances within the glomerular filtrate (such as glucose, amino acids, vitamins and water) are taken back into the blood after ultrafiltration.

Chapter 2. The reproductive and urinary systems

CHAPTER HIGHLIGHTS & NOTES: KEY TERMS, PEOPLE, PLACES, CONCEPTS

Secretion	Secretion is the process of elaborating, releasing, and oozing chemicals, or a secreted chemical substance from a cell or gland. In contrast to excretion, the substance may have a certain function, rather than being a waste product. Many cells contain this such as glucoma cells.
Ureter	In human anatomy, the ureters are muscular tubes that propel urine from the kidneys to the urinary bladder. In the adult, the ureters are usually 25-30 cm (10-12 in) long and ~3-4 mm in diameter. In humans, the ureters arise from the renal pelvis on the medial aspect of each kidney before descending towards the bladder on the front of the psoas major muscle. The ureters cross the pelvic brim near the bifurcation of the iliac arteries (which they run over). This is a common site for the impaction of kidney stones (the others being the ureterovesical valve and the pelviureteric junction where the ureter joins the renal pelvis in the renal hilum). The ureters run posteroinferiorly on the lateral walls of the pelvis and then curve anteriormedially to enter the bladder through the back, at the vesicoureteric junction, running within the wall of the bladder for a few centimetres. The backflow of urine is prevented by valves known as ureterovesical valves.
Circumcision	Male circumcision is the surgical removal of some or all of the foreskin (prepuce) from the penis. Early depictions of circumcision are found in cave paintings and Ancient Egyptian tombs, though some pictures are open to interpretation. Religious male circumcision is considered a commandment from God in Judaism.
Female genital mutilation	Terminology The procedures known as female\ genital\ mutilation were referred to as female circumcision until the early 1980s, when the term 'female genital mutilation' came into use. The term was adopted at the third conference of the Inter-African Committee on Traditional Practices Affecting the Health of Women and Children in Addis Ababa, Ethiopia, and in 1991 the WHO recommended its use to the United Nations. It has since become the dominant term within the international community and in medical literature.
Urinary tract infection	A urinary tract infection is a bacterial infection that affects part of the urinary tract. When it affects the lower urinary tract it is known as a simple cystitis (a bladder infection) and when it affects the upper urinary tract it is known as pyelonephritis (a kidney infection). Symptoms from a lower urinary tract include painful urination and either frequent urination or urge to urinate, while those of pyelonephritis include fever and flank pain in addition to the symptoms of a lower UTI. In the elderly and the very young, symptoms may be vague or non specific.
Urine	Urine is a typically sterile liquid by-product of the body secreted by the kidneys through a process called urination and excreted through the urethra.

Chapter 2. The reproductive and urinary systems

CHAPTER HIGHLIGHTS & NOTES: KEY TERMS, PEOPLE, PLACES, CONCEPTS

	Cellular metabolism generates numerous by-products, many rich in nitrogen, that require elimination from the bloodstream. These by-products are eventually expelled from the body in a process known as micturition, the primary method for excreting water-soluble chemicals from the body.
Chlamydia trachomatis	Chlamydia trachomatis, an obligate intracellular human pathogen, is one of three bacterial species in the genus Chlamydia. C. trachomatis is a Gram-negative bacteria, therefore its cell wall components retain the counter-stain safranin and appear pink under a light microscope. The inclusion bodies of Chlamydia trachomatis were first described in 1907, the Chlamydia trachomatis agent was first cultured in the yolk sacs of eggs by Feifan Tang et al in 1957.
Fallopian tube	The Fallopian Tube, uterine tubes, and salpinges are two very fine tubes lined with ciliated epithelia, leading from the ovaries of female mammals into the uterus, via the utero-tubal junction. In non-mammalian vertebrates, the equivalent structures are the oviducts. In a woman's body the tube allows passage of the egg from the ovary to the uterus.
Oviduct	In vertebrates, the passageway from the ovaries to the outside of the body is known as the oviduct. The eggs travel along the oviduct. These eggs will either be fertilized by sperm to become a zygote, or will degenerate in the body.
Uterus	The uterus is a major female hormone-responsive reproductive sex organ of most mammals including humans. One end, the cervix, opens into the vagina, while the other is connected to one or both fallopian tubes, depending on the species. It is within the uterus that the fetus develops during gestation, usually developing completely in placental mammals such as humans and partially in marsupials such as kangaroos and opossums.
Ligament	In anatomy, the term ligament is used to denote any of three types of structures. Most commonly, it refers to fibrous tissue that connects bones to other bones and is also known as articular ligament, articular larua, fibrous ligament, or true ligament. Ligament can also refer to:•Peritoneal ligament: a fold of peritoneum or other membranes.•Fetal remnant ligament: the remnants of a tubular structure from the fetal period of life. The study of ligaments is known as desmology .
Endometrium	The endometrium is the inner membrane of the mammalian uterus.

Chapter 2. The reproductive and urinary systems

CHAPTER HIGHLIGHTS & NOTES: KEY TERMS, PEOPLE, PLACES, CONCEPTS

	The endometrium is the innermost glandular layer and functions as a lining for the uterus, preventing adhesions between the opposed walls of the myometrium, thereby maintaining the patency of the uterine cavity. During the menstrual cycle or estrous cycle, the endometrium grows to a thick, blood vessel-rich, glandular tissue layer.
Methotrexate	Methotrexate ( ), abbreviated MTX and formerly known as amethopterin, is an antimetabolite and antifolate drug. It is used in treatment of cancer, autoimmune diseases, ectopic pregnancy, and for the induction of medical abortions. It acts by inhibiting the metabolism of folic acid.
Myometrium	The myometrium is the middle layer of the uterine wall, consisting mainly of uterine smooth muscle cells (also called uterine myocytes), but also of supporting stromal and vascular tissue. Its main function is to induce uterine contractions. The myometrium is located between the endometrium (the inner layer of the uterine wall), and the serosa or perimetrium (the outer uterine layer).
Salpingectomy	Salpingectomy refers to the surgical removal of a Fallopian tube. Indications The procedure was first performed by Lawson Tait in patients with a bleeding ectopic pregnancy; this procedure has since saved the lives of countless women. Other indications for a salpingectomy include infected tubes, (as in a hydrosalpinx) or as part of the surgical procedure for tubal cancer.
Ectopic pregnancy	An ectopic pregnancy, is a complication of pregnancy in which the embryo implants outside the uterine cavity. With rare exceptions, ectopic pregnancies are not viable. Furthermore, they are dangerous for the parent, since internal haemorrhage is a life threatening complication.
Gonadotropin	Glycoprotein hormone Gonadotropins are protein hormones secreted by gonadotrope cells of the pituitary gland of vertebrates. This is a family of proteins, which include the mammalian hormones follitropin (FSH), lutropin (LH), placental chorionic gonadotropins hCG and eCG and chorionic gonadotropin as well as at least two forms of fish gonadotropins. These hormones are central to the complex endocrine system that regulates normal growth, sexual development, and reproductive function.
Cervix	The cervix is the lower, narrow portion of the uterus where it joins with the top end of the vagina. It is cylindrical or conical in shape and protrudes through the upper anterior vaginal wall.

Chapter 2. The reproductive and urinary systems

CHAPTER HIGHLIGHTS & NOTES: KEY TERMS, PEOPLE, PLACES, CONCEPTS

Pancreas	The pancreas is a gland organ in the digestive and endocrine system of vertebrates. It is both an endocrine gland producing several important hormones, including insulin, glucagon, and somatostatin, and a digestive organ, secreting pancreatic juice containing digestive enzymes that assist the absorption of nutrients and the digestion in the small intestine. These enzymes help to further break down the carbohydrates, proteins, and lipids in the chyme.
Perimetrium	The perimetrium is the outer serosa layer of the uterus, equivalent to peritoneum.
Cancer	Cancer, known medically as a malignant neoplasm, is a broad group of various diseases, all involving unregulated cell growth. In cancer, cells divide and grow uncontrollably, forming malignant tumors, and invade nearby parts of the body. The cancer may also spread to more distant parts of the body through the lymphatic system or bloodstream.
Clitoris	The clitoris is a sexual organ that is present only in female mammals. In humans, the visible button-like portion is located near the anterior junction of the labia minora, above the opening of the urethra and vagina. Unlike the penis, which is homologous to the clitoris, the clitoris does not contain the distal portion of the urethra.
Hymen	The hymen is a membrane that surrounds or partially covers the external vaginal opening. It forms part of the vulva, or external genitalia. The size of the hymenal opening increases with age.
Labia majora	The labia majora are two prominent longitudinal cutaneous folds that extend downward and backward from the mons pubis to the perineum and form the lateral boundaries of the pudendal cleft, which contains the labia minora, interlabial sulci, clitoral hood, clitoral glans, frenulum clitoridis, the Hart's Line, and the vulval vestibule, which contains the external openings of the urethra and the vagina. Each labium majus has two surfaces, an outer, pigmented and covered with strong, crisp hairs; and an inner, smooth and beset with large sebaceous follicles. Between the two there is a considerable quantity of areolar tissue, fat, and a tissue resembling the dartos tunic of the scrotum, besides vessels, nerves, and glands.
Labia minora	The labia minora, inner lips, or nymphae, are two flaps of skin on either side of the human vaginal opening, situated between the labia majora (outer labia, or outer lips). Inner lips vary widely in size, colour, and shape from woman to woman. The inner lips extend from the clitoris obliquely downward, laterally, and backward on either side of the vulval vestibule, ending between the bottom of the vulval vestibule and the outer lips.
Vulva	The vulva consists of the external genital organs of the female mammal

Chapter 2. The reproductive and urinary systems

CHAPTER HIGHLIGHTS & NOTES: KEY TERMS, PEOPLE, PLACES, CONCEPTS

Pelvic floor	The pelvic floor is composed of muscle fibers of the levator ani, the coccygeus, and associated connective tissue which span the area underneath the pelvis. The pelvic diaphragm is a muscular partition formed by the levatores ani and coccygei, with which may be included the parietal pelvic fascia on their upper and lower aspects. The pelvic floor separates the pelvic cavity above from the perineal region (including perineum) below.
Pelvic inflammatory disease	Pelvic inflammatory disease is a generic term for inflammation of the uterus, fallopian tubes, and/or ovaries as it progresses to scar formation with adhesions to nearby tissues and organs. This may lead to infections. pelvic\ inflammatory\ disease is a vague term and can refer to viral, fungal, parasitic, though most often bacterial infections.
Cremaster muscle	The cremaster muscle is a muscle that covers the testis.
	Its function is to raise and lower the testes in order to regulate the temperature of the testes and promote spermatogenesis. Contraction may also occur during arousal which can prevent injury to the testicles during sex.
Scrotum	In some male mammals, the scrotum is a dual-chambered protuberance of skin and muscle, containing the testicles and divided by a septum. It is an extension of the perineum, and is located between the penis and anus. In humans and some other mammals, the scrotum becomes covered with curly pubic hairs at puberty.
Seminiferous tubule	Seminiferous tubules are located in the testes, and are the specific location of meiosis, and the subsequent creation of gametes, namely spermatozoa.
	The epithelium of the tubule consists of sustentacular or Sertoli cells, which are tall, columnar type cells that line the tubule.
	In between the Sertoli cells are spermatogenic cells, which differentiate through meiosis to sperm cells.
Tunica albuginea	Tunica albuginea is an anatomy term that literally means 'white covering.' It is used to refer to three anatomical areas which include:•Tunica albuginea the tough fibrous layer of connective tissue that surrounds the corpora cavernosa of the penis•Tunica albuginea a layer of connective tissue covering the testicles•Tunica albuginea the connective tissue covering of the ovaries.
Epididymis	The epididymis is part of the male reproductive system and is present in all male amniotes. It is a narrow, tightly-coiled tube connecting the efferent ducts from the rear of each testicle to its vas deferens. A similar, but probably non-homologous, structure is found in cartilaginous fishes.

Chapter 2. The reproductive and urinary systems

CHAPTER HIGHLIGHTS & NOTES: KEY TERMS, PEOPLE, PLACES, CONCEPTS

Vas deferens	The vas deferens, also called ductus deferens is part of the male anatomy of many vertebrates; they transport sperm from the epididymis in anticipation of ejaculation.
	There are two ducts, connecting the left and right epididymis to the ejaculatory ducts in order to move sperm. Each tube is about 30 centimeters long (in humans) and is muscular (surrounded by smooth muscle).
Ejaculatory duct	The ejaculatory ducts (ductus ejaculatorii) are paired structures in male anatomy. Each ejaculatory duct is formed by the union of the vas deferens with the duct of the seminal vesicle. They pass through the prostate, and open into the urethra at the Colliculus seminalis.
Foreskin	In male human anatomy, the foreskin is a generally retractable double-layered fold of skin and mucous membrane that covers the glans penis and protects the urinary meatus ( ) when the penis is not erect. It is also described as the prepuce, a technically broader term that also includes the clitoral hood in women, to which the foreskin is embryonically homologous.
	The outside of the foreskin is a continuation of the skin on the shaft of the penis, but the inner foreskin is a mucous membrane like the inside of the eyelid or the mouth.
Gamete	A gamete is a cell that fuses with another cell during fertilization (conception) in organisms that reproduce sexually. In species that produce two morphologically distinct types of gametes, and in which each individual produces only one type, a female is any individual that produces the larger type of gamete--called an ovum --and a male produces the smaller tadpole-like type--called a sperm. This is an example of anisogamy or heterogamy, the condition wherein females and males produce gametes of different sizes (this is the case in humans; the human ovum has approximately 100,000 times the volume of a single human sperm cell).
Gametogenesis	Gametogenesis is a biological process by which diploid or haploid precursor cells undergo cell division and differentiation to form mature haploid gametes. Depending on the biological life cycle of the organism, gametogenesis occurs by meiotic division of diploid gametocytes into various gametes, or by mitotic division of haploid gametogenous cells. For example, plants produce gametes through mitosis in gametophytes.
Penis	The penis is a biological feature of male animals including both vertebrates and invertebrates (creatures with and without backbones, respectively). It is a reproductive, intromittent organ that additionally serves as the urinal duct in placental mammals.
	The word 'penis' is taken from the Latin word for 'tail.' Some derive that from Indo-European *pesnis, and the Greek word π?ος = 'penis' from Indo-European *pesos.
Semen	Semen is an organic fluid, also known as seminal fluid, that may contain spermatozoa.

Chapter 2. The reproductive and urinary systems

CHAPTER HIGHLIGHTS & NOTES: KEY TERMS, PEOPLE, PLACES, CONCEPTS

	It is secreted by the gonads (sexual glands) and other sexual organs of male or hermaphroditic animals and can fertilize female ova. In humans, seminal fluid contains several components besides spermatozoa: proteolytic and other enzymes as well as fructose are elements of seminal fluid which promote the survival of spermatozoa and provide a medium through which they can move or 'swim'.
Seminal vesicle	The seminal vesicles (glandulae vesiculosae) or vesicular glands are a pair of simple tubular glands posteroinferior to the urinary bladder of male mammals. It is located within the pelvis. Each seminal gland spreads approximately 5 cm, though the full length of seminal vesicle is approximately 10 cm, but it is curled up inside of the gland's structure.
Follicle-stimulating hormone	Follicle-stimulating hormone is a hormone found in humans and other animals. It is synthesized and secreted by gonadotrophs of the anterior pituitary gland. FSH regulates the development, growth, pubertal maturation, and reproductive processes of the body.
Luteinizing hormone	Luteinizing hormone is a hormone produced by the anterior pituitary gland. In females, an acute rise of LH ('LH surge') triggers ovulation and development of the corpus luteum. In males, where LH had also been called interstitial cell-stimulating hormone (ICSH), it stimulates Leydig cell production of testosterone.
Mitosis	Mitosis is the process by which a eukaryotic cell separates the chromosomes in its cell nucleus into two identical sets in two nuclei. It is generally followed immediately by cytokinesis, which divides the nuclei, cytoplasm, organelles and cell membrane into two cells containing roughly equal shares of these cellular components. Mitosis and cytokinesis together define the mitotic (M) phase of the cell cycle--the division of the mother cell into two daughter cells, genetically identical to each other and to their parent cell.
Sertoli cell	A Sertoli cell is a 'nurse' cell of the testes that is part of a seminiferous tubule. It is activated by follicle-stimulating hormone and has FSH-receptor on its membranes. It is specifically located in the convoluted seminiferous tubules (since this is the only place in the testes where the spermatozoa is produced). Functions Because its main function is to nourish the developing sperm cells through the stages of spermatogenesis, the Sertoli cell has also been called the 'mother' or 'nurse' cell.
Spermatocyte	A spermatocyte is a male gametocyte, derived from a spermatogonium, which is in the developmental stage of spermatogenesis during which meiosis occurs. It is located in the seminiferous tubules of the testis.

Chapter 2. The reproductive and urinary systems

CHAPTER HIGHLIGHTS & NOTES: KEY TERMS, PEOPLE, PLACES, CONCEPTS

Spermatogenesis	Spermatogenesis is the process by which male primary sperm cells undergo meiosis, and produce a number of cells termed spermatogonia, from which the primary spermatocytes are derived. Each primary spermatocyte divides into two secondary spermatocytes, and each secondary spermatocyte into two spermatids or young spermatozoa. These develop into mature spermatozoa, also known as sperm cells.
Spermiogenesis	Spermiogenesis is the final stage of spermatogenesis, which sees the maturation of spermatids into mature, motile spermatozoa. The spermatid is more or less circular cell containing a nucleus, Golgi apparatus, centriole and mitochondria. All these components take part in forming the spermatozoon.
Testosterone	Testosterone is a steroid hormone from the androgen group and is found in mammals, reptiles, birds, and other vertebrates. In mammals, testosterone is primarily secreted in the testes of males and the ovaries of females, although small amounts are also secreted by the adrenal glands. It is the principal male sex hormone and an anabolic steroid.
Hypothalamus	The hypothalamus is a portion of the brain that contains a number of small nuclei with a variety of functions. One of the most important functions of the hypothalamus is to link the nervous system to the endocrine system via the pituitary gland (hypophysis). The hypothalamus is located below the thalamus, just above the brain stem.
Leydig cell	Leydig cells, also known as interstitial cells of Leydig, are found adjacent to the seminiferous tubules in the testicle. They produce testosterone in the presence of luteinizing hormone (LH). Leydig cells are polyhedral in shape, display a large prominent nucleus, an eosinophilic cytoplasm and numerous lipid-filled vesicles.

Cram101

Chapter 2. The reproductive and urinary systems

CHAPTER QUIZ: KEY TERMS, PEOPLE, PLACES, CONCEPTS

1. An _____, is a complication of pregnancy in which the embryo implants outside the uterine cavity. With rare exceptions, ectopic pregnancies are not viable. Furthermore, they are dangerous for the parent, since internal haemorrhage is a life threatening complication.

   a. Interstitial pregnancy
   b. Ovarian pregnancy
   c. Ectopic pregnancy
   d. Society of Obstetricians and Gynaecologists of Canada

2. Terminology

   The procedures known as female\ genital\ mutilation were referred to as female circumcision until the early 1980s, when the term '_____' came into use. The term was adopted at the third conference of the Inter-African Committee on Traditional Practices Affecting the Health of Women and Children in Addis Ababa, Ethiopia, and in 1991 the WHO recommended its use to the United Nations. It has since become the dominant term within the international community and in medical literature.

   a. Khalid Adem
   b. Female genital mutilation
   c. Female Genital Mutilation Act 2003
   d. Gishiri cutting

3. In human anatomy, the _____s are muscular tubes that propel urine from the kidneys to the urinary bladder. In the adult, the _____s are usually 25-30 cm (10-12 in) long and ~3-4 mm in diameter.

   In humans, the _____s arise from the renal pelvis on the medial aspect of each kidney before descending towards the bladder on the front of the psoas major muscle. The _____s cross the pelvic brim near the bifurcation of the iliac arteries (which they run over). This is a common site for the impaction of kidney stones (the others being the ureterovesical valve and the pelviureteric junction where the _____ joins the renal pelvis in the renal hilum). The _____s run posteroinferiorly on the lateral walls of the pelvis and then curve anteriormedially to enter the bladder through the back, at the vesicoureteric junction, running within the wall of the bladder for a few centimetres. The backflow of urine is prevented by valves known as ureterovesical valves.

   a. Izonsteride
   b. Ureter
   c. Bacillus anthracis phage AP50
   d. Bacterial cell structure

4. . _____ is the basic structural and functional unit of the kidney. Its chief function is to regulate the concentration of water and soluble substances like sodium salts by filtering the blood, reabsorbing what is needed and excreting the rest as urine. A _____ eliminates wastes from the body, regulates blood volume and blood pressure, controls levels of electrolytes and metabolites, and regulates blood pH. Its functions are vital to life and are regulated by the endocrine system by hormones such as antidiuretic hormone, aldosterone, and parathyroid hormone.

Chapter 2. The reproductive and urinary systems

CHAPTER QUIZ: KEY TERMS, PEOPLE, PLACES, CONCEPTS

a. Podocyte
b. Renal capsule
c. Nephron
d. Renal cortex

5. In chemistry, especially biochemistry, a _____ is a carboxylic acid with a long aliphatic tail (chain), which is either saturated or unsaturated. Most naturally occurring _____s have a chain of an even number of carbon atoms, from 4 to 28. _____s are usually derived from triglycerides or phospholipids. When they are not attached to other molecules, they are known as 'free' _____s.

a. Flavan-3-ol
b. Flavonoid
c. Fatty acid
d. Food Balance Wheel

ANSWER KEY
Chapter 2. The reproductive and urinary systems

1. c
2. b
3. b
4. c
5. c

You can take the complete Chapter Practice Test

for Chapter 2. The reproductive and urinary systems
on all key terms, persons, places, and concepts.

Online 99 Cents

http://www.epub5042.32.20612.2.cram101.com/

Use www.Cram101.com for all your study needs

including Cram101's online interactive problem solving labs in

chemistry, statistics, mathematics, and more.

Chapter 3. Endocrinology

CHAPTER OUTLINE: KEY TERMS, PEOPLE, PLACES, CONCEPTS

	Endocrine system
	Hypothalamus
	Ectopic pregnancy
	Endocrinology
	Gonadotropin
	Nervous system
	Secretion
	Attenuated vaccine
	Cholesterol
	Glucocorticoid
	Menstrual cycle
	Mineralocorticoid
	Neurotransmitter
	Oxytocin
	Pheromone
	Pregnenolone
	Sex steroid
	Steroid hormone
	Testosterone

Chapter 3. Endocrinology
CHAPTER OUTLINE: KEY TERMS, PEOPLE, PLACES, CONCEPTS

	Thyroid hormone
	Anabolic steroid
	Arachidonic acid
	Follicle-stimulating hormone
	Growth hormone
	Human placental lactogen
	Leukotriene
	Luteinizing hormone
	Peptide hormone
	Phospholipase C
	Prolactin
	Prostaglandin
	Thyroid-stimulating hormone
	Hormone
	Placental lactogen
	Dopamine
	Melatonin
	Prohormone
	Tyrosine

Chapter 3. Endocrinology

CHAPTER OUTLINE: KEY TERMS, PEOPLE, PLACES, CONCEPTS

_____ Vasoactive intestinal peptide

_____ Activation

_____ Homeostasis

_____ Necrotizing enterocolitis

_____ Enterocolitis

_____ Agonist

_____ Antagonist

_____ Blood cell

_____ Gland

_____ Reproduction

_____ Adrenocorticotropic hormone

_____ Calcium

_____ Hypothyroidism

_____ Parathyroid gland

_____ Parathyroid hormone

_____ Pituitary gland

_____ Thyroxine

_____ Triiodothyronine

_____ Adrenal gland

Chapter 3. Endocrinology
CHAPTER OUTLINE: KEY TERMS, PEOPLE, PLACES, CONCEPTS

	Aldosterone
	Corticosteroid
	Gonad
	Carbohydrate metabolism
	Metabolism

CHAPTER HIGHLIGHTS & NOTES: KEY TERMS, PEOPLE, PLACES, CONCEPTS

Endocrine system	The endocrine system is the system of glands, each of which secretes a type of hormone directly into the bloodstream to regulate the body. The endocrine system is in contrast to the exocrine system, which secretes its chemicals using ducts. It derives from the Greek words 'endo' meaning inside, within, and 'crinis' for secrete.
Hypothalamus	The hypothalamus is a portion of the brain that contains a number of small nuclei with a variety of functions. One of the most important functions of the hypothalamus is to link the nervous system to the endocrine system via the pituitary gland (hypophysis). The hypothalamus is located below the thalamus, just above the brain stem.
Ectopic pregnancy	An ectopic pregnancy, is a complication of pregnancy in which the embryo implants outside the uterine cavity. With rare exceptions, ectopic pregnancies are not viable. Furthermore, they are dangerous for the parent, since internal haemorrhage is a life threatening complication.
Endocrinology	Endocrinology is a branch of biology and medicine dealing with the endocrine system, its diseases, and its specific secretions called hormones, the integration of developmental events such as proliferation, growth, and differentiation (including histogenesis and organogenesis) and the coordination of metabolism, respiration, excretion, movement, reproduction, and sensory perception depend on chemical cues, substances synthesized and secreted by specialized cells.

Chapter 3. Endocrinology

CHAPTER HIGHLIGHTS & NOTES: KEY TERMS, PEOPLE, PLACES, CONCEPTS

Endocrinology is concerned with the study of the biosynthesis, storage, chemistry, and physiological function of hormones and with the cells of the endocrine glands and tissues that secrete them.

The endocrine system consists of several glands, all and in different parts of the body, that secrete hormones directly into the blood rather than into a duct system.

Gonadotropin	Glycoprotein hormone
	Gonadotropins are protein hormones secreted by gonadotrope cells of the pituitary gland of vertebrates. This is a family of proteins, which include the mammalian hormones follitropin (FSH), lutropin (LH), placental chorionic gonadotropins hCG and eCG and chorionic gonadotropin as well as at least two forms of fish gonadotropins. These hormones are central to the complex endocrine system that regulates normal growth, sexual development, and reproductive function.
Nervous system	The nervous system is an organ system containing a network of specialized cells called neurons that coordinate the actions of an animal and transmit signals between different parts of its body. In most animals the nervous system consists of two parts, central and peripheral. The central nervous system of vertebrates (such as humans) contains the brain, spinal cord, and retina. The peripheral nervous system consists of sensory neurons, clusters of neurons called ganglia, and nerves connecting them to each other and to the central nervous system. These regions are all interconnected by means of complex neural pathways. The enteric nervous system, a subsystem of the peripheral nervous system, has the capacity, even when severed from the rest of the nervous system through its primary connection by the vagus nerve, to function independently in controlling the gastrointestinal system.
Secretion	Secretion is the process of elaborating, releasing, and oozing chemicals, or a secreted chemical substance from a cell or gland. In contrast to excretion, the substance may have a certain function, rather than being a waste product. Many cells contain this such as glucoma cells.
Attenuated vaccine	An attenuated vaccine is a vaccine created by reducing the virulence of a pathogen, but still keeping it viable . Attenuation takes an infectious agent and alters it so that it becomes harmless or less virulent. These vaccines contrast to those produced by 'killing' the virus (inactivated vaccine).
Cholesterol	Cholesterol is an organic chemical substance classified as a waxy steroid of fat. It is an essential structural component of mammalian cell membranes and is required to establish proper membrane permeability and fluidity.

Chapter 3. Endocrinology

CHAPTER HIGHLIGHTS & NOTES: KEY TERMS, PEOPLE, PLACES, CONCEPTS

	In addition to its importance within cells, cholesterol is an important component in the hormonal systems of the body for the manufacture of bile acids, steroid hormones, and vitamin D. Cholesterol is the principal sterol synthesized by animals; in vertebrates it is formed predominantly in the liver.
Glucocorticoid	Glucocorticoids are a class of steroid hormones that bind to the glucocorticoid receptor (GR), which is present in almost every vertebrate animal cell.
	Glucocorticoids are part of the feedback mechanism in the immune system that turns immune activity (inflammation) down. They are therefore used in medicine to treat diseases that are caused by an overactive immune system, such as allergies, asthma, autoimmune diseases and sepsis. Glucocorticoids have many diverse (pleiotropic) effects, including potentially harmful side effects, and as a result are rarely sold over-the-counter. They also interfere with some of the abnormal mechanisms in cancer cells, so they are used in high doses to treat cancer.
Menstrual cycle	The menstrual cycle is the scientific term for the physiological changes that can occur in fertile women for the purposes of sexual reproduction and fertilization
	The menstrual cycle, under the control of the endocrine system, is necessary for reproduction.
Mineralocorticoid	Mineralocorticoids are a class of steroid hormones characterised by their similarity to aldosterone and their influence on salt and water balances. The primary endogenous mineralocorticoid is aldosterone, although a number of other endogenous hormones (including progesterone and deoxycorticosterone) have mineralocorticoid function.
	Aldosterone acts on the kidneys to provide active reabsorption of sodium and an associated passive reabsorption of water, as well as the active secretion of potassium in the principal cells of the cortical collecting tubule and active secretion of protons via proton ATPases in the lumenal membrane of the intercalated cells of the collecting tubule.
Neurotransmitter	Neurotransmitters are endogenous chemicals which transmit signals from a neuron to a target cell across a synapse. Neurotransmitters are packaged into synaptic vesicles clustered beneath the membrane on the presynaptic side of a synapse, and are released into the synaptic cleft, where they bind to receptors in the membrane on the postsynaptic side of the synapse. Release of neurotransmitters usually follows arrival of an action potential at the synapse, but may also follow graded electrical potentials.
Oxytocin	Oxytocin ( ) is a mammalian hormone that acts primarily as a neuromodulator in the brain.

Chapter 3. Endocrinology

CHAPTER HIGHLIGHTS & NOTES: KEY TERMS, PEOPLE, PLACES, CONCEPTS

	Oxytocin is best known for its roles in sexual reproduction, in particular during and after childbirth. It is released in large amounts after distension of the cervix and uterus during labor, facilitating birth, and after stimulation of the nipples, facilitating breastfeeding.
Pheromone	A pheromone is a secreted or excreted chemical factor that triggers a social response in members of the same species. Pheromones are chemicals capable of acting outside the body of the secreting individual to impact the behavior of the receiving individual. There are alarm pheromones, food trail pheromones, sex pheromones, and many others that affect behavior or physiology.
Pregnenolone	Pregnenolone, also known as 3α,5β-tetrahydroprogesterone (3α,5β-THP), is an endogenous steroid hormone involved in the steroidogenesis of progestogens, mineralocorticoids, glucocorticoids, androgens, and estrogens, as well as the neuroactive steroids. As such it is a prohormone, though it also has biological effects of its own, behaving namely as a neuroactive steroid in its own right with potent anxiolytic effects. Like other steroids, pregnenolone consists of four interconnected cyclic hydrocarbons.
Sex steroid	Sex steroids, also known as gonadal steroids, are steroid hormones that interact with vertebrate androgen or estrogen receptors. Their effects are mediated by slow genomic mechanisms through nuclear receptors as well as by fast nongenomic mechanisms through membrane-associated receptors and signaling cascades. The term sex hormone is nearly always synonymous with sex steroid.
Steroid hormone	A steroid hormone is a steroid that acts as a hormone. Steroid hormones can be grouped into five groups by the receptors to which they bind: glucocorticoids, mineralocorticoids, androgens, estrogens, and progestogens. Vitamin D derivatives are a sixth closely related hormone system with homologous receptors, though technically sterols rather than steroids.
Testosterone	Testosterone is a steroid hormone from the androgen group and is found in mammals, reptiles, birds, and other vertebrates. In mammals, testosterone is primarily secreted in the testes of males and the ovaries of females, although small amounts are also secreted by the adrenal glands. It is the principal male sex hormone and an anabolic steroid.
Thyroid hormone	The thyroid hormones, thyroxine (T_4) and triiodothyronine (T_3), are tyrosine-based hormones produced by the thyroid gland primarily responsible for regulation of metabolism. An important component in the synthesis of thyroid hormones is iodine. The major form of thyroid hormone in the blood is thyroxine (T_4), which has a longer half life than T_3.

Chapter 3. Endocrinology

CHAPTER HIGHLIGHTS & NOTES: KEY TERMS, PEOPLE, PLACES, CONCEPTS

Anabolic steroid	Anabolic steroids, technically known as anabolic-androgen steroids (AAS) or colloquially as 'steroids' (or even 'roids'), are drugs that mimic the effects of testosterone and dihydrotestosterone in the body. They increase protein synthesis within cells, which results in the buildup of cellular tissue (anabolism), especially in muscles. Anabolic steroids also have androgenic and virilizing properties, including the development and maintenance of masculine characteristics such as the growth of the vocal cords, testicles, and body hair (secondary sexual characteristics).
Arachidonic acid	Arachidonic acid is a polyunsaturated omega-6 fatty acid $20:4(\omega-6)$. It is the counterpart to the saturated arachidic acid found in peanut oil, (L. arachis - peanut).
	Arachidonic acid is one of the essential fatty acids required by most mammals. Some mammals lack the ability to--or have a very limited capacity to--convert linoleic acid into arachidonic acid, making it an essential part of their diet. Since little or no arachidonic acid is found in common plants, such animals are obligate carnivores; the cat is a common example. A commercial source of arachidonic acid has been derived, however, from the fungus Mortierella alpina
Follicle-stimulating hormone	Follicle-stimulating hormone is a hormone found in humans and other animals. It is synthesized and secreted by gonadotrophs of the anterior pituitary gland. FSH regulates the development, growth, pubertal maturation, and reproductive processes of the body.
Growth hormone	Growth hormone is a peptide hormone that stimulates growth, cell reproduction and regeneration in humans and other animals. Growth hormone is a 191-amino acid, single-chain polypeptide that is synthesized, stored, and secreted by the somatotroph cells within the lateral wings of the anterior pituitary gland. Somatotropin (STH) refers to the growth hormone 1 produced naturally in animals, whereas the term somatropin refers to growth hormone produced by recombinant DNA technology, and is abbreviated 'HGH' in humans.
Human placental lactogen	Human placental lactogen also called human chorionic somatomammotropin, is a polypeptide placental hormone. Its structure and function is similar to that of human growth hormone. It modifies the metabolic state of the mother during pregnancy to facilitate the energy supply of the fetus.
Leukotriene	Leukotrienes are fatty signalling molecules. They were first found in leukocytes (hence their name). One of their roles (specifically, leukotriene D_4) is to trigger contractions in the smooth muscles lining the trachea; their overproduction is a major cause of inflammation in asthma and allergic rhinitis. Leukotriene antagonists are used to treat these diseases by inhibiting the production or activity of leukotrienes.
Luteinizing hormone	Luteinizing hormone is a hormone produced by the anterior pituitary gland. In females, an acute rise of LH ('LH surge') triggers ovulation and development of the corpus luteum.

Chapter 3. Endocrinology

CHAPTER HIGHLIGHTS & NOTES: KEY TERMS, PEOPLE, PLACES, CONCEPTS

Peptide hormone	Peptide hormones are a class of peptides that are secreted into the blood stream and have endocrine functions in living animals. Like other proteins, peptide hormones are synthesized in cells from amino acids according to an mRNA template, which is itself synthesized from a DNA template inside the cell nucleus. Peptide hormone precursors (pre-prohormones) are then processed in several stages, typically in the endoplasmic reticulum, including removal of the N-terminal signal sequence and sometimes glycosylation, resulting in prohormones. The prohormones are then packaged into membrane-bound secretory vesicles, which can be secreted from the cell by exocytosis in response to specific stimuli e.g increase of calcium and cAMP concentration in cytoplasm.
Phospholipase C	Phospholipase C is a class of enzymes that cleave phospholipids just before the phosphate group. It is most commonly taken to be synonymous with the human forms of this enzyme, which plays an important role in eukaryotic cell physiology, in particular signal transduction pathways. Thirteen kinds of mammalian phospholipase C are classified into six models (β, γ, δ, ε, ζ, η) according to structure.
Prolactin	Prolactin also known as luteotropic hormone (LTH) is a protein that in humans is encoded by the PRL gene. Prolactin is a peptide hormone discovered by Henry Friesen. Although it is perhaps best known for its role in lactation, prolactin already existed in the oldest known vertebrates--fish--where its most important functions were probably related to control of water and salt balance.
Prostaglandin	A prostaglandin is any member of a group of lipid compounds that are derived enzymatically from fatty acids and have important functions in the animal body. Every prostaglandin contains 20 carbon atoms, including a 5-carbon ring. They are mediators and have a variety of strong physiological effects, such as regulating the contraction and relaxation of smooth muscle tissue. Prostaglandins are not hormones, but autocrine or paracrine, which are locally acting messenger molecules. They differ from hormones in that they are not produced at a discrete site but in many places throughout the human body. Also, their target cells are present in the immediate vicinity of the site of their secretion (of which there are many).
Thyroid-stimulating hormone	Thyroid-stimulating hormone is a hormone that stimulates the thyroid gland to produce thyroxine (T_4), and then triiodothyronine (T_3) which stimulates the metabolism of almost every tissue in the body. It is a glycoprotein hormone synthesized and secreted by thyrotrope cells in the anterior pituitary gland, which regulates the endocrine function of the thyroid gland. Regulation of thyroid hormone levels

Chapter 3. Endocrinology

CHAPTER HIGHLIGHTS & NOTES: KEY TERMS, PEOPLE, PLACES, CONCEPTS

Hormone	A hormone is a chemical released by a cell or a gland in one part of the body that sends out messages that affect cells in other parts of the organism. Only a small amount of hormone is required to alter cell metabolism. In essence, it is a chemical messenger that transports a signal from one cell to another.
Placental lactogen	Placental lactogen is a polypeptide placental hormone. Its structure and function is similar to that of growth hormone. It modifies the metabolic state of the mother during pregnancy to facilitate the energy supply of the fetus.
Dopamine	Dopamine is a simple organic chemical in the catecholamine family, is a monoamine neurotransmitter which plays a number of important physiological roles in the bodies of animals. In addition to being a catecholamine and a monoamine, dopamine may be classified as a substituted phenethylamine. Its name derives from its chemical structure, which consists of an amine group (NH_2) linked to a catechol structure called dihydroxyphenethylamine, the decarboxylated form of dihydroxyphenylalanine (acronym DOPA).
Melatonin	Melatonin, also known chemically as N-acetyl-5-methoxytryptamine, is a naturally occurring compound found in animals, plants, and microbes. In animals, circulating levels of the hormone melatonin vary in a daily cycle, thereby allowing the entrainment of the circadian rhythms of several biological functions. Many biological effects of melatonin are produced through activation of melatonin receptors, while others are due to its role as a pervasive and powerful antioxidant, with a particular role in the protection of nuclear and mitochondrial DNA. Products containing melatonin have been available over-the-counter in the United States since the mid-1990s.
Prohormone	A prohormone refers to a committed intra-glandular precursor of a hormone, usually having minimal hormonal effect by itself. The term has been used in medical science since the middle of the 20th century. The primary function of a prohormone is to enhance the strength of the hormone that already occurs in the body.
Tyrosine	Tyrosine is one of the 20 amino acids that are used by cells to synthesize proteins. Its codons are UAC and UAU. It is a non-essential amino acid with a polar side group. The word 'tyrosine' is from the Greek tyri, meaning cheese, as it was first discovered in 1846 by German chemist Justus von Liebig in the protein casein from cheese.
Vasoactive intestinal peptide	Vasoactive intestinal peptide is produced in many tissues of vertebrates including the gut, pancreas and suprachiasmatic nuclei of the hypothalamus in the brain.

Chapter 3. Endocrinology

CHAPTER HIGHLIGHTS & NOTES: KEY TERMS, PEOPLE, PLACES, CONCEPTS

	VIP stimulates contractility in the heart, causes vasodilation, increases glycogenolysis, lowers arterial blood pressure and relaxes the smooth muscle of trachea, stomach and gall bladder. In humans, the vasoactive intestinal peptide is encoded by the VIP gene.
Activation	Activation in (bio-)chemical sciences generally refers to the process whereby something is prepared or excited for a subsequent reaction.
	In chemistry, activation of molecules is where the molecules enter a state that avails for a chemical reaction to occur. The phrase energy of activation refers to the energy the reactants must acquire before they can successfully react with each other to produce the products, that is, to reach the transition state. The energy needed for activation can be quite small and the molecules may have enough energy just from thermal fluctuations the molecules naturally have (i.e. lots of reactions don't have to be heated to proceed). The branch of chemistry that deals with this topic is called chemical kinetics.
Homeostasis	Homeostasis is the property of a system that regulates its internal environment and tends to maintain a stable, constant condition of properties such as temperature or pH. It can be either an open or closed system.
	It was defined by Claude Bernard and later by Walter Bradford Cannon in 1926, 1929 and 1932.
	Typically used to refer to a living organism, the concept came from that of milieu intérieur that was created by Claude Bernard and published in 1865. Multiple dynamic equilibrium adjustment and regulation mechanisms make homeostasis possible.
Necrotizing enterocolitis	Necrotizing enterocolitis is a medical condition primarily seen in premature infants, where portions of the bowel undergo necrosis (tissue death).
	The condition is typically seen in premature infants, and the timing of its onset is generally inversely proportional to the gestational age of the baby at birth, i.e. the earlier a baby is born, the later signs of NEC are typically seen. Initial symptoms include feeding intolerance, increased gastric residuals, abdominal distension and bloody stools.
Enterocolitis	Enterocolitis is an inflammation of the colon and small intestine.
Agonist	An agonist is a chemical that binds to a receptor of a cell and triggers a response by that cell. Agonists often mimic the action of a naturally occurring substance. Whereas an agonist causes an action, an antagonist blocks the action of the agonist and an inverse agonist causes an action opposite to that of the agonist.

Chapter 3. Endocrinology

CHAPTER HIGHLIGHTS & NOTES: KEY TERMS, PEOPLE, PLACES, CONCEPTS

Antagonist	Most muscles work in pairs, and when a muscle works it needs to have an agonist and an antagonist, unless the muscle's natural state is opposite to that which is produced by the muscle, example Sphincter ani externus muscle. An 'antagonist' is a classification used to describe a muscle that acts in opposition to the specific movement generated by the agonist and is responsible for returning a limb to its initial position. Antagonistic Pairs Antagonistic muscles are found in pairs called antagonistic pairs.
Blood cell	A blood cell, is a cell produced by haematopoiesis and normally found in blood. In mammals, these fall into three general categories:•Red blood cells -- Erythrocytes•White blood cells -- Leukocytes•Platelets -- Thrombocytes. Together, these three kinds of blood cells add up to a total 45% of the blood tissue by volume, with the remaining 55% of the volume composed of plasma, the liquid component of blood. This volume percentage (e.g., 45%) of cells to total volume is called hematocrit, determined by centrifuge or flow cytometry.
Gland	A gland is an organ in an animal's body that synthesizes a substance for release of substances such as hormones or breast milk, often into the bloodstream (endocrine gland) or into cavities inside the body or its outer surface (exocrine gland). Glands can be divided into 2 groups:•Endocrine glands -- are glands that secrete their products through the basal lamina and lack a duct system.•Exocrine glands -- secrete their products through a duct or directly onto the apical surface, the glands in this group can be divided into three groups: •Apocrine glands -- a portion of the secreting cell's body is lost during secretion. Apocrine gland is often used to refer to the apocrine sweat glands, however it is thought that apocrine sweat glands may not be true apocrine glands as they may not use the apocrine method of secretion.•Holocrine glands -- the entire cell disintegrates to secrete its substances (e.g., sebaceous glands)•Merocrine glands -- cells secrete their substances by exocytosis (e.g., mucous and serous glands).
Reproduction	Reproduction is the biological process by which new 'offspring' individual organisms are produced from their 'parents'. Reproduction is a fundamental feature of all known life; each individual organism exists as the result of reproduction. The known methods of reproduction are broadly grouped into two main types: sexual and asexual.

Chapter 3. Endocrinology

CHAPTER HIGHLIGHTS & NOTES: KEY TERMS, PEOPLE, PLACES, CONCEPTS

Adrenocorticotropic hormone	Adrenocorticotropic hormone also known as corticotropin, is a polypeptide tropic hormone produced and secreted by the anterior pituitary gland. It is an important component of the hypothalamic-pituitary-adrenal axis and is often produced in response to biological stress (along with its precursor corticotropin-releasing hormone from the hypothalamus). Its principal effects are increased production and release of corticosteroids.
Calcium	Calcium is the chemical element with the symbol Ca and atomic number 20. It has an atomic mass of 40.078 amu. Calcium is a soft gray alkaline earth metal, and is the fifth-most-abundant element by mass in the Earth's crust. Calcium is also the fifth-most-abundant dissolved ion in seawater by both molarity and mass, after sodium, chloride, magnesium, and sulfate.
Hypothyroidism	Hypothyroidism is a condition in which the thyroid gland does not make enough thyroid hormone. Iodine deficiency is often cited as the most common cause of hypothyroidism worldwide but it can be caused by many other factors. It can result from a lack of a thyroid gland or from iodine-131 treatment, and can also be associated with increased stress.
Parathyroid gland	The parathyroid glands are small endocrine glands in the neck that produce parathyroid hormone. Humans usually have four parathyroid glands, which are usually located on the rear surface of the thyroid gland, or, in rare cases, within the thyroid gland itself or in the chest. Parathyroid glands control the amount of calcium in the blood and within the bones.
Parathyroid hormone	Parathyroid hormone parathormone or parathyrin, is secreted by the parathyroid glands as a polypeptide containing 84 amino acids. It acts to increase the concentration of calcium (Ca^{2+}) in the blood, whereas calcitonin (a hormone produced by the parafollicular cells (C cells) of the thyroid gland) acts to decrease calcium concentration. PTH acts to increase the concentration of calcium in the blood by acting upon parathyroid hormone receptor in three parts of the body: PTH half-life is approximately 4 minutes.
Pituitary gland	In vertebrate anatomy the pituitary gland, is an endocrine gland about the size of a pea and weighing 0.5 grams (0.018 oz) in humans. It is not a part of the brain. It is a protrusion off the bottom of the hypothalamus at the base of the brain, and rests in a small, bony cavity (sella turcica) covered by a dural fold (diaphragma sellae).
Thyroxine	Thyroxine, or 3,5,3',5'-tetraiodothyronine , a form of thyroid hormones is the major hormone secreted by the follicular cells of the thyroid gland. Synthesis and regulation

Chapter 3. Endocrinology

CHAPTER HIGHLIGHTS & NOTES: KEY TERMS, PEOPLE, PLACES, CONCEPTS

	Thyroxine is synthesized via the iodination and covalent bonding of the phenyl portions of tyrosine residues found in an initial peptide, thyroglobulin, which is secreted into thyroid granules. These iodinated diphenyl compounds are cleaved from their peptide backbone upon being stimulated by thyroid-stimulating hormone.
Triiodothyronine	Triiodothyronine, $C_{15}H_{12}I_3NO_4$, also known as T_3, is a thyroid hormone. It affects almost every physiological process in the body, including growth and development, metabolism, body temperature, and heart rate.
Adrenal gland	In mammals, the adrenal glands (also known as suprarenal glands) are endocrine glands that sit atop the kidneys; in humans, the right adrenal gland is triangular shaped, while the left adrenal gland is semilunar shaped. They are chiefly responsible for releasing hormones in response to stress through the synthesis of corticosteroids such as cortisol and catecholamines such as epinephrine (adrenaline) and norepinephrine. They also produce androgens.
Aldosterone	Aldosterone is a yellow steroid hormone (mineralocorticoid family) produced by the outer-section (zona glomerulosa) of the adrenal cortex in the adrenal gland, and acts mainly on the distal tubules and collecting ducts of the nephron, the functioning unit of the kidney, to cause the conservation of sodium, secretion of potassium, increased water retention, and increased blood pressure. The overall effect of aldosterone is to increase reabsorption of ions and water in the kidney -- increasing blood volume and, therefore, increasing blood pressure. Drugs that interfere with the secretion or action of aldosterone are in use as antihypertensives.
Corticosteroid	Corticosteroids are a class of chemicals that includes steroid hormones naturally produced in the adrenal cortex of vertebrates and analogues of these hormones that are synthesized in laboratories. Corticosteroids are involved in a wide range of physiologic processes, including stress response, immune response, and regulation of inflammation, carbohydrate metabolism, protein catabolism, blood electrolyte levels, and behavior. •Glucocorticoids such as cortisol control carbohydrate, fat and protein metabolism and are anti-inflammatory by preventing phospholipid release, decreasing eosinophil action and a number of other mechanisms. •Mineralocorticoids such as aldosterone control electrolyte and water levels, mainly by promoting sodium retention in the kidney. Some common natural hormones are corticosterone ($C_{21}H_{30}O_4$), cortisone ($C_{21}H_{28}O_5$, 17-hydroxy-11-dehydrocorticosterone) and aldosterone.
Gonad	The gonad is the organ that makes gametes. The gonads in males are the testes and the gonads in females are the ovaries.

Chapter 3. Endocrinology

CHAPTER HIGHLIGHTS & NOTES: KEY TERMS, PEOPLE, PLACES, CONCEPTS

Carbohydrate metabolism	Carbohydrate metabolism denotes the various biochemical processes responsible for the formation, breakdown and interconversion of carbohydrates in living organisms. The most important carbohydrate is glucose, a simple sugar (monosaccharide) that is metabolized by nearly all known organisms. Glucose and other carbohydrates are part of a wide variety of metabolic pathways across species: plants synthesize carbohydrates from atmospheric gases by photosynthesis storing the absorbed energy internally, often in the form of starch or lipids.
Metabolism	Metabolism is the set of chemical reactions that happen in the cells of living organisms to sustain life. These processes allow organisms to grow and reproduce, maintain their structures, and respond to their environments. The word metabolism can also refer to all chemical reactions that occur in living organisms, including digestion and the transport of substances into and between different cells, in which case the set of reactions within the cells is called intermediary metabolism or intermediate metabolism.

CHAPTER QUIZ: KEY TERMS, PEOPLE, PLACES, CONCEPTS

1. The _____ is the scientific term for the physiological changes that can occur in fertile women for the purposes of sexual reproduction and fertilization

   The _____, under the control of the endocrine system, is necessary for reproduction.

   a. Menstrual leave
   b. Chhaupadi
   c. Concealed ovulation
   d. Menstrual cycle

2. The _____s, thyroxine (T_4) and triiodothyronine (T_3), are tyrosine-based hormones produced by the thyroid gland primarily responsible for regulation of metabolism. An important component in the synthesis of _____s is iodine. The major form of _____ in the blood is thyroxine (T_4), which has a longer half life than T_3.

   a. Thyroid hormone
   b. Thyroxine
   c. magnetic resonance imaging
   d. Pregnenolone

3. . _____ is the process of elaborating, releasing, and oozing chemicals, or a secreted chemical substance from a cell or gland. In contrast to excretion, the substance may have a certain function, rather than being a waste product.

Chapter 3. Endocrinology

CHAPTER QUIZ: KEY TERMS, PEOPLE, PLACES, CONCEPTS

Many cells contain this such as glucoma cells.

a. Secretion
b. N-Acyl homoserine lactone
c. Bacillus anthracis phage AP50
d. Bacterial cell structure

4. An _____, is a complication of pregnancy in which the embryo implants outside the uterine cavity. With rare exceptions, ectopic pregnancies are not viable. Furthermore, they are dangerous for the parent, since internal haemorrhage is a life threatening complication.

a. Ectopic pregnancy
b. Ovarian pregnancy
c. Adrenalectomy
d. Adrenarche

5. _____s are endogenous chemicals which transmit signals from a neuron to a target cell across a synapse. _____s are packaged into synaptic vesicles clustered beneath the membrane on the presynaptic side of a synapse, and are released into the synaptic cleft, where they bind to receptors in the membrane on the postsynaptic side of the synapse. Release of _____s usually follows arrival of an action potential at the synapse, but may also follow graded electrical potentials.

a. Superior laryngeal nerve
b. magnetic resonance imaging
c. Neurotransmitter
d. Cortisone

ANSWER KEY
Chapter 3. Endocrinology

1. d
2. a
3. a
4. a
5. c

You can take the complete Chapter Practice Test

for Chapter 3. Endocrinology
on all key terms, persons, places, and concepts.

Online 99 Cents

http://www.epub5042.32.20612.3.cram101.com/

Use www.Cram101.com for all your study needs

including Cram101's online interactive problem solving labs in

chemistry, statistics, mathematics, and more.

Chapter 4. Reproductive cycles

CHAPTER OUTLINE: KEY TERMS, PEOPLE, PLACES, CONCEPTS

_____ Folic acid

_____ Follicle-stimulating hormone

_____ Luteinizing hormone

_____ Osteoclast

_____ Follicular phase

_____ Stromal cell

_____ Zona pellucida

_____ Interleukin

_____ Growth factor

_____ Kidney

_____ Pituitary gland

_____ Renin-angiotensin system

_____ Corpus luteum

_____ Luteal phase

_____ Ovulation

_____ Polar body

_____ Volume

_____ Secretion

_____ Endometrium

Chapter 4. Reproductive cycles
CHAPTER OUTLINE: KEY TERMS, PEOPLE, PLACES, CONCEPTS

_____ Fallopian tube

_____ Myometrium

_____ Natural family planning

_____ Family planning

_____ Progesterone

_____ Pheromone

_____ Premenstrual syndrome

_____ Menstruation

_____ Neuropeptide Y

_____ Infertility

_____ Combined oral contraceptive pill

_____ Hyperprolactinaemia

_____ Obesity

_____ Polycystic ovary syndrome

_____ Emergency contraception

_____ Mifepristone

_____ Menarche

_____ Norethisterone

_____ Puberty

Chapter 4. Reproductive cycles

CHAPTER OUTLINE: KEY TERMS, PEOPLE, PLACES, CONCEPTS

| Precocious puberty
| Delayed puberty
| Growth hormone
| Grandmother hypothesis
| Menopause
| Endocrinology
| Reproductive system
| Hormone replacement therapy
| Osteoblast

CHAPTER HIGHLIGHTS & NOTES: KEY TERMS, PEOPLE, PLACES, CONCEPTS

Folic acid	Folic acid and folate (the form naturally occurring in the body), as well as pteroyl-L-glutamic acid, pteroyl-L-glutamate, and pteroylmonoglutamic acid are forms of the water-soluble vitamin B_9. Folic acid is itself not biologically active, but its biological importance is due to tetrahydrofolate and other derivatives after its conversion to dihydrofolic acid in the liver. Vitamin B_9 (folic acid and folate inclusive) is essential to numerous bodily functions.
Follicle-stimulating hormone	Follicle-stimulating hormone is a hormone found in humans and other animals. It is synthesized and secreted by gonadotrophs of the anterior pituitary gland. FSH regulates the development, growth, pubertal maturation, and reproductive processes of the body.
Luteinizing hormone	Luteinizing hormone is a hormone produced by the anterior pituitary gland. In females, an acute rise of LH ('LH surge') triggers ovulation and development of the corpus luteum.

Chapter 4. Reproductive cycles

CHAPTER HIGHLIGHTS & NOTES: KEY TERMS, PEOPLE, PLACES, CONCEPTS

Osteoclast	An osteoclast (from the Greek words for 'bone' and 'broken') is a type of bone cell that removes bone tissue by removing its mineralized matrix and breaking up the organic bone (organic dry weight is 90% collagen). This process is known as bone resorption. Osteoclasts were discovered by Kolliker in 1873. Osteoclasts and osteoblasts are instrumental in controlling the amount of bone tissue: osteoblasts form bone, osteoclasts resorb bone.
Follicular phase	The follicular phase is the phase of the estrous cycle, (or, in humans and great apes, the menstrual cycle) during which follicles in the ovary mature. It ends with ovulation. The main hormone controlling this stage is estradiol.
Stromal cell	In cell biology, stromal cells are connective tissue cells of an organ found in the loose connective tissue. These are most often associated with the uterine mucosa (endometrium), prostate, bone marrow precursor cells, and the ovary as well as the hematopoietic system and elsewhere. These are the cells that make up the support structure of biological tissues and support the parenchymal cells.
Zona pellucida	The zona pellucida is a glycoprotein membrane surrounding the plasma membrane of an oocyte. It is a vital constitutive part of the oocyte, external but of essential importance to it. The zona pellucida first appears in multilaminar primary oocytes.
Interleukin	Interleukins are a group of cytokines (secreted proteinssignaling molecules) that were first seen to be expressed by white blood cells (leukocytes). The term interleukin derives from (inter-) 'as a means of communication', and (-leukin) 'deriving from the fact that many of these proteins are produced by leukocytes and act on leukocytes'. The name is something of a relic, though (the term was coined by Dr. Vern Paetkau, University of Victoria); it has since been found that interleukins are produced by a wide variety of body cells.
Growth factor	A growth factor is a naturally occurring substance capable of stimulating cellular growth, proliferation and cellular differentiation. Usually it is a protein or a steroid hormone. Growth factors are important for regulating a variety of cellular processes.
Kidney	The kidneys are organs with several functions. They are seen in many types of animals, including vertebrates and some invertebrates. They are an essential part of the urinary system and also serve homeostatic functions such as the regulation of electrolytes, maintenance of acid-base balance, and regulation of blood pressure. They serve the body as a natural filter of the blood, and remove wastes which are diverted to the urinary bladder. In producing urine, the kidneys excrete wastes such as urea and ammonium; the kidneys also are responsible for the reabsorption of water, glucose, and amino acids. The kidneys also produce hormones including calcitriol, renin, and erythropoietin.

Chapter 4. Reproductive cycles

CHAPTER HIGHLIGHTS & NOTES: KEY TERMS, PEOPLE, PLACES, CONCEPTS

Pituitary gland	In vertebrate anatomy the pituitary gland, is an endocrine gland about the size of a pea and weighing 0.5 grams (0.018 oz) in humans. It is not a part of the brain. It is a protrusion off the bottom of the hypothalamus at the base of the brain, and rests in a small, bony cavity (sella turcica) covered by a dural fold (diaphragma sellae).
Renin-angiotensin system	The renin-angiotensin system or the renin-angiotensin-aldosterone system (RAAS) is a hormone system that regulates blood pressure and water (fluid) balance. When blood volume is low, juxtaglomerular cells in the kidneys secrete renin. Renin stimulates the production of angiotensin I, which is then converted to angiotensin II. Angiotensin II causes blood vessels to constrict, resulting in increased blood pressure.
Corpus luteum	The corpus luteum is a temporary endocrine structure in female mammals, involved in production of relatively high levels of progesterone and moderate levels of estradiol and inhibin A. It is colored as a result of concentrating carotenoids from the diet. Corpus luteum secretes moderate amount of estrogen to inhibit further release of GnRH and thus secretion of LH and FSH. The corpus luteum develops from an ovarian follicle during the luteal phase of the menstrual cycle or estrous cycle, following the release of a secondary oocyte from the follicle during ovulation. The follicle first forms a corpus hemorrhagicum before it becomes a corpus luteum, but the term refers to the visible collection of blood left after rupture of the follicle that secretes progesterone.
Luteal phase	The luteal phase is the latter phase of the menstrual cycle (in humans and a few other animals) or the estrous cycle (in other placental mammals). It begins with the formation of the corpus luteum and ends in either pregnancy or luteolysis. The main hormone associated with this stage is progesterone, which is significantly higher during the luteal phase than other phases of the cycle.
Ovulation	Ovulation is the process in a female's menstrual cycle by which a mature ovarian follicle ruptures and discharges an ovum (also known as an oocyte, female gamete, or casually, an egg). Ovulation also occurs in the estrous cycle of other female mammals, which differs in many fundamental ways from the menstrual cycle. The time immediately surrounding ovulation is referred to as the ovulatory phase or the periovulatory period.
Polar body	A polar body is a cell structure found inside an ovum. Both animal and plant ova possess it. It is also known as a polar cell.

Chapter 4. Reproductive cycles

CHAPTER HIGHLIGHTS & NOTES: KEY TERMS, PEOPLE, PLACES, CONCEPTS

Volume	Volume is how much three-dimensional space a substance (solid, liquid, gas, or plasma) or shape occupies or contains, often quantified numerically using the SI derived unit, the cubic metre. The volume of a container is generally understood to be the capacity of the container, i. e. the amount of fluid (gas or liquid) that the container could hold, rather than the amount of space the container itself displaces.
Secretion	Secretion is the process of elaborating, releasing, and oozing chemicals, or a secreted chemical substance from a cell or gland. In contrast to excretion, the substance may have a certain function, rather than being a waste product. Many cells contain this such as glucoma cells.
Endometrium	The endometrium is the inner membrane of the mammalian uterus.
	The endometrium is the innermost glandular layer and functions as a lining for the uterus, preventing adhesions between the opposed walls of the myometrium, thereby maintaining the patency of the uterine cavity. During the menstrual cycle or estrous cycle, the endometrium grows to a thick, blood vessel-rich, glandular tissue layer.
Fallopian tube	The Fallopian Tube, uterine tubes, and salpinges are two very fine tubes lined with ciliated epithelia, leading from the ovaries of female mammals into the uterus, via the utero-tubal junction. In non-mammalian vertebrates, the equivalent structures are the oviducts.
	In a woman's body the tube allows passage of the egg from the ovary to the uterus.
Myometrium	The myometrium is the middle layer of the uterine wall, consisting mainly of uterine smooth muscle cells (also called uterine myocytes), but also of supporting stromal and vascular tissue. Its main function is to induce uterine contractions.
	The myometrium is located between the endometrium (the inner layer of the uterine wall), and the serosa or perimetrium (the outer uterine layer).
Natural family planning	Natural family planning is a term referring to the family planning methods approved by the Roman Catholic Church. In accordance with the Church's teachings regarding sexual behavior in keeping with its philosophy of the dignity of the human person, natural\ family\ planning excludes the use of other methods of birth control, which it refers to as 'artificial contraception.'
	Periodic abstinence is the only method deemed moral by the Church for avoiding pregnancy. When used to avoid pregnancy, natural\ family\ planning limits sexual intercourse to naturally infertile periods; portions of the menstrual cycle, during pregnancy, and after menopause.

Chapter 4. Reproductive cycles

CHAPTER HIGHLIGHTS & NOTES: KEY TERMS, PEOPLE, PLACES, CONCEPTS

Family planning	Family planning is the planning of when to have children, and the use of birth control and other techniques to implement such plans. Other techniques commonly used include sexuality education, prevention and management of sexually transmitted infections, pre-conception counseling and management, and infertility management. Family planning is sometimes used in the wrong way also as a synonym for the use of birth control, though it often includes more.
Progesterone	Progesterone also known as P4 (pregn-4-ene-3,20-dione) is a C-21 steroid hormone involved in the female menstrual cycle, pregnancy (supports gestation) and embryogenesis of humans and other species. Progesterone belongs to a class of hormones called progestogens, and is the major naturally occurring human progestogen. Progesterone was independently discovered by four research groups.
Pheromone	A pheromone is a secreted or excreted chemical factor that triggers a social response in members of the same species. Pheromones are chemicals capable of acting outside the body of the secreting individual to impact the behavior of the receiving individual. There are alarm pheromones, food trail pheromones, sex pheromones, and many others that affect behavior or physiology.
Premenstrual syndrome	Premenstrual syndrome (also called PMT or premenstrual tension) is a collection of physical and emotional symptoms related to a woman's menstrual cycle. While most women of child-bearing age (up to 85%) report having experienced physical symptoms related to normal ovulatory function, such as bloating or breast tenderness, medical definitions of PMS are limited to a consistent pattern of emotional and physical symptoms occurring only during the luteal phase of the menstrual cycle that are of 'sufficient severity to interfere with some aspects of life'. In particular, emotional symptoms must be present consistently to diagnose PMS. The specific emotional and physical symptoms attributable to PMS vary from woman to woman, but each individual woman's pattern of symptoms is predictable, occurs consistently during the ten days prior to menses, and vanishes either shortly before or shortly after the start of menstrual flow.
Menstruation	Menstruation is the shedding of the uterine lining (endometrium).
Neuropeptide Y	Neuropeptide Y is a 36-amino acid peptide neurotransmitter found in the brain and autonomic nervous system. It regulates energy usage, and is involved in learning, memory processing, and epilepsy. The main effect of its level/activity elevation is increased food intake and decreased physical activity.
Infertility	Infertility primarily refers to the biological inability of a person to contribute to conception. Infertility may also refer to the state of a woman who is unable to carry a pregnancy to full term.

Chapter 4. Reproductive cycles

CHAPTER HIGHLIGHTS & NOTES: KEY TERMS, PEOPLE, PLACES, CONCEPTS

Combined oral contraceptive pill	The combined oral contraceptive pill often referred to as the birth-control pill or colloquially as 'the Pill', is a birth control method that includes a combination of an estrogen (oestrogen) and a progestin (progestogen). When taken by mouth every day, these pills inhibit female fertility. They were first approved for contraceptive use in the United States in the 1950s, and are a very popular form of birth control.
Hyperprolactinaemia	Hyperprolactinaemia is the presence of abnormally-high levels of prolactin in the blood. Normal levels are less than 500 mIU/L for women, and less than 450 mIU/L for men. Prolactin is a peptide hormone produced by the anterior pituitary gland primarily associated with lactation and plays a vital role in breast development during pregnancy.
Obesity	Obesity is a medical condition in which excess body fat has accumulated to the extent that it may have an adverse effect on health, leading to reduced life expectancy and/or increased health problems. Body mass index (BMI), a measurement which compares weight and height, defines people as overweight (pre-obese) if their BMI is between 25 and 30 kg/m^2, and obese when it is greater than 30 kg/m^2. Obesity increases the likelihood of various diseases, particularly heart disease, type 2 diabetes, obstructive sleep apnea, certain types of cancer, and osteoarthritis.
Polycystic ovary syndrome	Polycystic ovary syndrome is one of the most common female endocrine disorders. PCOS is a complex, heterogeneous disorder of uncertain etiology, but there is strong evidence that it can to a large degree be classified as a genetic disease. PCOS produces symptoms in approximately 5% to 10% of women of reproductive age (12-45 years old).
Emergency contraception	Emergency contraception or emergency postcoital contraception, are birth control measures that, if taken after sexual intercourse, may prevent pregnancy. Forms of emergency\ contraception include:•Emergency contraceptive pills (ECPs)--sometimes simply referred to as emergency contraceptives or the 'morning-after pill'--are drugs intended to disrupt ovulation or fertilization, which are steps necessary for pregnancy (contraceptives). There is controversy about whether such drugs may in some cases act not as a contraceptive but as a contragestive, a drug that prevents the implantation of a human embryo in the uterus, thus preventing pregnancy, although one study has concluded that this mechanism is unlikely.•Intrauterine devices (IUDs)--usually used as a primary contraceptive method, but sometimes used as emergency contraception.Emergency contraceptive pills (ECPs)

Chapter 4. Reproductive cycles

CHAPTER HIGHLIGHTS & NOTES: KEY TERMS, PEOPLE, PLACES, CONCEPTS

Mifepristone	Mifepristone is a synthetic steroid compound used as a pharmaceutical. It is a progesterone receptor antagonist used as an abortifacient in the first months of pregnancy, and in smaller doses as an emergency contraceptive. Mifepristone is also a powerful glucocorticoid receptor antagonist, and has occasionally been used in refractory Cushing's Syndrome (due to ectopic/neoplastic ACTH/Cortisol secretion).
Menarche	Menarche is the first menstrual cycle, or first menstrual bleeding, in female human beings. From both social and medical perspectives it is often considered the central event of female puberty, as it signals the possibility of fertility. Girls experience menarche at different ages.
Norethisterone	Norethisterone (or 19-nor-17α-ethynyltestosterone) is a molecule used in some combined oral contraceptive pills, progestogen only pills and is also available as a stand-alone drug. It is a progestogen and can be used to treat premenstrual syndrome, painful periods, abnormal heavy bleeding, irregular periods, menopausal syndrome (in combination with oestrogen), or to postpone a period. It is also commonly used to help prevent uterine hemorrhage in complicated non-surgical or pre-surgical gynecologic cases.
Puberty	Puberty is the process of physical changes by which a child's body matures into an adult body capable of sexual reproduction to enable fertilisation. It is initiated by hormonal signals from the brain to the gonads; the ovaries in a girl, the testes in a boy. In response to the signals, the gonads produce hormones that stimulate libido and the growth, function, and transformation of the brain, bones, muscle, blood, skin, hair, breasts, and sexual organs.
Precocious puberty	As a medical term, precocious puberty describes puberty occurring at an unusually early age. In most of these children, the process is normal in every respect except the unusually early age, and simply represents a variation of normal development. In a minority of children, the early development is triggered by a disease such as a tumor or injury of the brain.
Delayed puberty	Puberty is described as delayed puberty with exceptions when an organism has passed the usual age of onset of puberty with no physical or hormonal signs that it is beginning. Puberty may be delayed for several years and still occur normally, in which case it is considered constitutional delay, a variation of healthy physical development. Delay of puberty may also occur due to malnutrition, many forms of systemic disease, or to defects of the reproductive system (hypogonadism) or the body's responsiveness to sex hormones.
Growth hormone	Growth hormone is a peptide hormone that stimulates growth, cell reproduction and regeneration in humans and other animals. Growth hormone is a 191-amino acid, single-chain polypeptide that is synthesized, stored, and secreted by the somatotroph cells within the lateral wings of the anterior pituitary gland.

Chapter 4. Reproductive cycles
CHAPTER HIGHLIGHTS & NOTES: KEY TERMS, PEOPLE, PLACES, CONCEPTS

Grandmother hypothesis	The grandmother hypothesis is a theory to explain the existence of menopause, rare in mammal species, in human life history and how a long post-fertile period (up to one third of a woman's lifespan) could confer an evolutionary advantage.
	Holding longevity constant, a female that undergoes menopause should have a lower total fertility rate, making menopause intriguing from an evolutionary perspective. Background
	In female placentals, the number of ovarian oocytes is fixed during embryonic development; possibly as an adaptation to reduce the accumulation of mutations.
Menopause	Menopause is a term used to describe the permanent cessation of the primary functions of the human ovaries: the ripening and release of ova and the release of hormones that cause both the creation of the uterine lining and the subsequent shedding of the uterine lining (a.k.a. the menses or the period). Menopause typically (but not always) occurs in women in midlife, during their late 40s or early 50s, and signals the end of the fertile phase of a woman's life.
	The transition from reproductive to non-reproductive is the result of a reduction in female hormonal production by the ovaries.
Endocrinology	Endocrinology is a branch of biology and medicine dealing with the endocrine system, its diseases, and its specific secretions called hormones, the integration of developmental events such as proliferation, growth, and differentiation (including histogenesis and organogenesis) and the coordination of metabolism, respiration, excretion, movement, reproduction, and sensory perception depend on chemical cues, substances synthesized and secreted by specialized cells.
	Endocrinology is concerned with the study of the biosynthesis, storage, chemistry, and physiological function of hormones and with the cells of the endocrine glands and tissues that secrete them.
	The endocrine system consists of several glands, all and in different parts of the body, that secrete hormones directly into the blood rather than into a duct system.
Reproductive system	The reproductive system is a system of organs within an organism which work together for the purpose of reproduction. Many non-living substances such as fluids, hormones, and pheromones are also important accessories to the reproductive system. Unlike most organ systems, the sexes of differentiated species often have significant differences.
Hormone replacement therapy	Hormone replacement therapy (HRT) is a system of medical treatment for surgically menopausal, transgender, perimenopausal and postmenopausal women.

Chapter 4. Reproductive cycles

CHAPTER HIGHLIGHTS & NOTES: KEY TERMS, PEOPLE, PLACES, CONCEPTS

	It is based on the idea that the treatment may prevent discomfort caused by diminished circulating oestrogen and progesterone hormones, and in the case of the surgically or prematurely menopausal, that it may prolong life and may reduce incidence of dementia. It involves the use of one or more of a group of medications designed to artificially boost hormone levels.
Osteoblast	Osteoblasts are mononucleate cells that are responsible for bone formation; in essence, osteoblasts are specialized fibroblasts that in addition to fibroblastic products, express bone sialoprotein and osteocalcin. Osteoblasts produce a matrix of osteoid, which is composed mainly of Type I collagen. Osteoblasts are also responsible for mineralization of this matrix.

CHAPTER QUIZ: KEY TERMS, PEOPLE, PLACES, CONCEPTS

1. _____ is a hormone produced by the anterior pituitary gland. In females, an acute rise of LH ('LH surge') triggers ovulation and development of the corpus luteum. In males, where LH had also been called interstitial cell-stimulating hormone (ICSH), it stimulates Leydig cell production of testosterone.

    a. Melanocyte-stimulating hormone
    b. Prolactin
    c. Luteinizing hormone
    d. Tropic hormone

2. _____ is one of the most common female endocrine disorders. PCOS is a complex, heterogeneous disorder of uncertain etiology, but there is strong evidence that it can to a large degree be classified as a genetic disease.

    PCOS produces symptoms in approximately 5% to 10% of women of reproductive age (12-45 years old).

    a. magnetic resonance imaging
    b. Adiponectin
    c. Polycystic ovary syndrome
    d. Agouti-related peptide

3. . _____ is a term referring to the family planning methods approved by the Roman Catholic Church. In accordance with the Church's teachings regarding sexual behavior in keeping with its philosophy of the dignity of the human person, natural\ family\ planning excludes the use of other methods of birth control, which it refers to as 'artificial contraception.'

    Periodic abstinence is the only method deemed moral by the Church for avoiding pregnancy.

Chapter 4. Reproductive cycles

CHAPTER QUIZ: KEY TERMS, PEOPLE, PLACES, CONCEPTS

When used to avoid pregnancy, natural\ family\ planning limits sexual intercourse to naturally infertile periods; portions of the menstrual cycle, during pregnancy, and after menopause.

a. Toni Weschler
b. Parametrium
c. Peg cell
d. Natural family planning

4. _____ (also called PMT or premenstrual tension) is a collection of physical and emotional symptoms related to a woman's menstrual cycle. While most women of child-bearing age (up to 85%) report having experienced physical symptoms related to normal ovulatory function, such as bloating or breast tenderness, medical definitions of PMS are limited to a consistent pattern of emotional and physical symptoms occurring only during the luteal phase of the menstrual cycle that are of 'sufficient severity to interfere with some aspects of life'. In particular, emotional symptoms must be present consistently to diagnose PMS. The specific emotional and physical symptoms attributable to PMS vary from woman to woman, but each individual woman's pattern of symptoms is predictable, occurs consistently during the ten days prior to menses, and vanishes either shortly before or shortly after the start of menstrual flow.

a. Ritushuddhi
b. Sanitary napkin
c. Seclusion of girls at puberty
d. Premenstrual syndrome

5. _____ is the first menstrual cycle, or first menstrual bleeding, in female human beings. From both social and medical perspectives it is often considered the central event of female puberty, as it signals the possibility of fertility.

Girls experience _____ at different ages.

a. Menarche
b. Menstrual taboo
c. Menstruation
d. Niddah

ANSWER KEY
Chapter 4. Reproductive cycles

1. c
2. c
3. d
4. d
5. a

You can take the complete Chapter Practice Test

for Chapter 4. Reproductive cycles
on all key terms, persons, places, and concepts.

Online 99 Cents

http://www.epub5042.32.20612.4.cram101.com/

Use www.Cram101.com for all your study needs

including Cram101's online interactive problem solving labs in

chemistry, statistics, mathematics, and more.

Chapter 5. Sexual differentiation and behaviour

CHAPTER OUTLINE: KEY TERMS, PEOPLE, PLACES, CONCEPTS

	Sexual differentiation
	Ejaculatory duct
	Y chromosome
	Testosterone
	Cryptorchidism
	Endocrine system
	Menopause
	Congenital adrenal hyperplasia
	Adrenal hyperplasia
	Sex steroid

CHAPTER HIGHLIGHTS & NOTES: KEY TERMS, PEOPLE, PLACES, CONCEPTS

Sexual differentiation	Sexual differentiation is the process of development of the differences between males and females from an undifferentiated zygote (fertilized egg). As male and female individuals develop from zygotes into fetuses, into infants, children, adolescents, and eventually into adults, sex and gender differences at many levels develop: genes, chromosomes, gonads, hormones, anatomy, psyche, and social behaviors. Sex differences range from nearly absolute to simply statistical.
Ejaculatory duct	The ejaculatory ducts (ductus ejaculatorii) are paired structures in male anatomy. Each ejaculatory duct is formed by the union of the vas deferens with the duct of the seminal vesicle. They pass through the prostate, and open into the urethra at the Colliculus seminalis.

Chapter 5. Sexual differentiation and behaviour

CHAPTER HIGHLIGHTS & NOTES: KEY TERMS, PEOPLE, PLACES, CONCEPTS

Y chromosome	The Y chromosome is one of the two sex-determining chromosomes in most mammals, including humans. In mammals, it contains the gene SRY, which triggers testis development if present. The human Y chromosome is composed of about 50 million base pairs.
Testosterone	Testosterone is a steroid hormone from the androgen group and is found in mammals, reptiles, birds, and other vertebrates. In mammals, testosterone is primarily secreted in the testes of males and the ovaries of females, although small amounts are also secreted by the adrenal glands. It is the principal male sex hormone and an anabolic steroid.
Cryptorchidism	Cryptorchidism is the absence of one or both testes from the scrotum. It is the most common birth defect regarding male genitalia. In unique cases, cryptorchidism can develop later in life, often as late as young adulthood.
Endocrine system	The endocrine system is the system of glands, each of which secretes a type of hormone directly into the bloodstream to regulate the body. The endocrine system is in contrast to the exocrine system, which secretes its chemicals using ducts. It derives from the Greek words 'endo' meaning inside, within, and 'crinis' for secrete.
Menopause	Menopause is a term used to describe the permanent cessation of the primary functions of the human ovaries: the ripening and release of ova and the release of hormones that cause both the creation of the uterine lining and the subsequent shedding of the uterine lining (a.k.a. the menses or the period). Menopause typically (but not always) occurs in women in midlife, during their late 40s or early 50s, and signals the end of the fertile phase of a woman's life. The transition from reproductive to non-reproductive is the result of a reduction in female hormonal production by the ovaries.
Congenital adrenal hyperplasia	Congenital adrenal hyperplasia refers to any of several autosomal recessive diseases resulting from mutations of genes for enzymes mediating the biochemical steps of production of cortisol from cholesterol by the adrenal glands (steroidogenesis). Most of these conditions involve excessive or deficient production of sex steroids and can alter development of primary or secondary sex characteristics in some affected infants, children, or adults. Associated conditions The symptoms of CAH vary depending upon the form of CAH and the gender of the patient.
Adrenal hyperplasia	Congenital adrenal hyperplasia refers to any of several autosomal recessive diseases resulting from mutations of genes for enzymes mediating the biochemical steps of production of cortisol from cholesterol by the adrenal glands (steroidogenesis).

Chapter 5. Sexual differentiation and behaviour

CHAPTER HIGHLIGHTS & NOTES: KEY TERMS, PEOPLE, PLACES, CONCEPTS

	Most of these conditions involve excessive or deficient production of sex steroids and can alter development of primary or secondary sex characteristics in some affected infants, children, or adults. Associated conditions
	The symptoms of CAH vary depending upon the form of CAH and the gender of the patient.
Sex steroid	Sex steroids, also known as gonadal steroids, are steroid hormones that interact with vertebrate androgen or estrogen receptors. Their effects are mediated by slow genomic mechanisms through nuclear receptors as well as by fast nongenomic mechanisms through membrane-associated receptors and signaling cascades. The term sex hormone is nearly always synonymous with sex steroid.

CHAPTER QUIZ: KEY TERMS, PEOPLE, PLACES, CONCEPTS

1. _____s, also known as gonadal steroids, are steroid hormones that interact with vertebrate androgen or estrogen receptors. Their effects are mediated by slow genomic mechanisms through nuclear receptors as well as by fast nongenomic mechanisms through membrane-associated receptors and signaling cascades. The term sex hormone is nearly always synonymous with _____.

    a. Transition nuclear protein
    b. Sex steroid
    c. Vagina
    d. Zona pellucida

2. _____ is the process of development of the differences between males and females from an undifferentiated zygote (fertilized egg). As male and female individuals develop from zygotes into fetuses, into infants, children, adolescents, and eventually into adults, sex and gender differences at many levels develop: genes, chromosomes, gonads, hormones, anatomy, psyche, and social behaviors.

    Sex differences range from nearly absolute to simply statistical.

    a. Sexual penetration
    b. Transvestophilia
    c. Venus Butterfly
    d. Sexual differentiation

3. . The _____s (ductus ejaculatorii) are paired structures in male anatomy.

Chapter 5. Sexual differentiation and behaviour

CHAPTER QUIZ: KEY TERMS, PEOPLE, PLACES, CONCEPTS

Each _____ is formed by the union of the vas deferens with the duct of the seminal vesicle. They pass through the prostate, and open into the urethra at the Colliculus seminalis.

a. Excretory duct of seminal gland
b. External spermatic fascia
c. External urethral orifice
d. Ejaculatory duct

4. _____ is a steroid hormone from the androgen group and is found in mammals, reptiles, birds, and other vertebrates. In mammals, _____ is primarily secreted in the testes of males and the ovaries of females, although small amounts are also secreted by the adrenal glands. It is the principal male sex hormone and an anabolic steroid.

a. Vasopressin
b. Pituitary gland
c. Pregnancy fetishism
d. Testosterone

5. The _____ is one of the two sex-determining chromosomes in most mammals, including humans. In mammals, it contains the gene SRY, which triggers testis development if present. The human _____ is composed of about 50 million base pairs.

a. Y chromosome microdeletion
b. External spermatic fascia
c. Y chromosome
d. Internal spermatic fascia

ANSWER KEY
Chapter 5. Sexual differentiation and behaviour

1. b
2. d
3. d
4. d
5. c

You can take the complete Chapter Practice Test

for Chapter 5. Sexual differentiation and behaviour

on all key terms, persons, places, and concepts.

Online 99 Cents

http://www.epub5042.32.20612.5.cram101.com/

Use www.Cram101.com for all your study needs

including Cram101's online interactive problem solving labs in

chemistry, statistics, mathematics, and more.

Chapter 6. Fertilization

CHAPTER OUTLINE: KEY TERMS, PEOPLE, PLACES, CONCEPTS

	Erectile dysfunction
	Erection
	Tadalafil
	Vardenafil
	Guanosine monophosphate
	Ejaculation
	Orgasm
	Oxytocin
	Premature ejaculation
	Gamete
	Osteoclast
	Y chromosome
	Ovulation
	Spermiogenesis
	Motility
	Hamster zona-free ovum test
	Zona pellucida
	Acrosome reaction
	Cortical reaction

Chapter 6. Fertilization
CHAPTER OUTLINE: KEY TERMS, PEOPLE, PLACES, CONCEPTS

_____ Mitosis

_____ Ubiquitin

_____ Sperm

_____ Hydatidiform mole

_____ Decidualization

_____ Endometrium

_____ Ectopic pregnancy

_____ Fetus

_____ Gonadotropin

_____ Immune system

_____ Fallopian tube

_____ Pelvic inflammatory disease

_____ Polycystic ovary syndrome

_____ Prostaglandin

_____ Female infertility

_____ Male infertility

_____ Platelet-activating factor

_____ Assisted reproductive technology

_____ Reproductive technology

Chapter 6. Fertilization

CHAPTER OUTLINE: KEY TERMS, PEOPLE, PLACES, CONCEPTS

	Angelman syndrome
	Ejaculatory duct
	Percutaneous epididymal sperm aspiration

CHAPTER HIGHLIGHTS & NOTES: KEY TERMS, PEOPLE, PLACES, CONCEPTS

Erectile dysfunction	Erectile dysfunction is sexual dysfunction characterized by the inability to develop or maintain an erection of the penis during sexual performance. A penile erection is the hydraulic effect of blood entering and being retained in sponge-like bodies within the penis. The process is often initiated as a result of sexual arousal, when signals are transmitted from the brain to nerves in the penis.
Erection	Penile erection is a physiological phenomenon where the penis becomes enlarged and firm. Penile erection is the result of a complex interaction of psychological, neural, vascular and endocrine factors, and is usually, though not exclusively, associated with sexual arousal or sexual attraction. The angle of an erect penis varies from pointing downwards, upwards, sideways, or may bend.
Tadalafil	Tadalafil is a PDE5 inhibitor, currently marketed in pill form for treating erectile dysfunction (ED) under the name Cialis; and under the name Adcirca for the treatment of pulmonary arterial hypertension. It initially was developed by the biotechnology company ICOS, and then again developed and marketed world-wide by Lilly ICOS, LLC, the joint venture of ICOS Corporation and Eli Lilly and Company. Cialis tablets, in 5 mg, 10 mg, and 20 mg doses, are yellow, film-coated, and almond-shaped.
Vardenafil	Vardenafil is a PDE5 inhibitor used for treating erectile dysfunction that is sold under the trade names Levitra (Bayer AG, GSK, and SP) and Staxyn. Vardenafil was co-marketed by Bayer Pharmaceuticals, GlaxoSmithKline, and Schering-Plough under the trade name Levitra. As of 2005, the co-promotion rights of GSK on Levitra have been returned to Bayer in many markets outside the U.S.

Chapter 6. Fertilization

CHAPTER HIGHLIGHTS & NOTES: KEY TERMS, PEOPLE, PLACES, CONCEPTS

Guanosine monophosphate	Guanosine monophosphate is a nucleotide that is found in RNA. It is an ester of phosphoric acid with the nucleoside guanosine. Guanosine monophosphate consists of the phosphate group, the pentose sugar ribose, and the nucleobase guanine; hence it is a ribonucleoside monophosphate. Guanosine monophosphate is produced from dried fish or dried seaweed.
Ejaculation	Ejaculation is the ejection of semen (usually carrying sperm) from the male reproductive tract, and is usually accompanied by orgasm. It is usually the final stage and natural objective of male sexual stimulation, and an essential component of natural conception. In rare cases ejaculation occurs because of prostatic disease.
Orgasm	Orgasm is the sudden discharge of accumulated sexual tension during the sexual response cycle, resulting in rhythmic muscular contractions in the pelvic region characterized by an intense sensation of pleasure. Experienced by males and females, orgasms are controlled by the involuntary, or autonomic, limbic system. They are often associated with other involuntary actions, including muscular spasms in multiple areas of the body, a general euphoric sensation and, frequently, body movements and vocalizations are expressed.
Oxytocin	Oxytocin ( ) is a mammalian hormone that acts primarily as a neuromodulator in the brain. Oxytocin is best known for its roles in sexual reproduction, in particular during and after childbirth. It is released in large amounts after distension of the cervix and uterus during labor, facilitating birth, and after stimulation of the nipples, facilitating breastfeeding.
Premature ejaculation	Premature ejaculation is a condition in which a man ejaculates earlier than he or his partner would like him to. Premature ejaculation is also known as rapid ejaculation, rapid climax, premature climax, or early ejaculation. Masters and Johnson defines PE as the condition in which a man ejaculates before his sex partner achieves orgasm, in more than fifty percent of their sexual encounters.
Gamete	A gamete is a cell that fuses with another cell during fertilization (conception) in organisms that reproduce sexually. In species that produce two morphologically distinct types of gametes, and in which each individual produces only one type, a female is any individual that produces the larger type of gamete--called an ovum --and a male produces the smaller tadpole-like type--called a sperm. This is an example of anisogamy or heterogamy, the condition wherein females and males produce gametes of different sizes (this is the case in humans; the human ovum has approximately 100,000 times the volume of a single human sperm cell).
Osteoclast	An osteoclast (from the Greek words for 'bone' and 'broken') is a type of bone cell that removes bone tissue by removing its mineralized matrix and breaking up the organic bone (organic dry weight is 90% collagen). This process is known as bone resorption. Osteoclasts were discovered by Kolliker in 1873.

Chapter 6. Fertilization

CHAPTER HIGHLIGHTS & NOTES: KEY TERMS, PEOPLE, PLACES, CONCEPTS

Y chromosome	The Y chromosome is one of the two sex-determining chromosomes in most mammals, including humans. In mammals, it contains the gene SRY, which triggers testis development if present. The human Y chromosome is composed of about 50 million base pairs.
Ovulation	Ovulation is the process in a female's menstrual cycle by which a mature ovarian follicle ruptures and discharges an ovum (also known as an oocyte, female gamete, or casually, an egg). Ovulation also occurs in the estrous cycle of other female mammals, which differs in many fundamental ways from the menstrual cycle. The time immediately surrounding ovulation is referred to as the ovulatory phase or the periovulatory period.
Spermiogenesis	Spermiogenesis is the final stage of spermatogenesis, which sees the maturation of spermatids into mature, motile spermatozoa. The spermatid is more or less circular cell containing a nucleus, Golgi apparatus, centriole and mitochondria. All these components take part in forming the spermatozoon.
Motility	Motility is a biological term which refers to the ability to move spontaneously and actively, consuming energy in the process. Most animals are motile but the term applies to unicellular and simple multicellular organisms, as well as to some mechanisms of fluid flow in multicellular organs, in addition to animal locomotion. Motile marine animals are commonly called free-swimming.
Hamster zona-free ovum test	The hamster zona-free ovum test or hamster test is a method for diagnosing male infertility due to the inability of the sperm to penetrate the ova. This test has limited value for most people experiencing infertility. In this test, sperm are incubated with several hamster eggs.
Zona pellucida	The zona pellucida is a glycoprotein membrane surrounding the plasma membrane of an oocyte. It is a vital constitutive part of the oocyte, external but of essential importance to it. The zona pellucida first appears in multilaminar primary oocytes.
Acrosome reaction	During fertilization, a sperm must first fuse with the plasma membrane and then penetrate the female egg in order to fertilize it. Fusing to the egg usually causes little problem, whereas penetrating through the egg's hard shell can present more of a problem to the sperm. Therefore sperm cells go through a process known as the acrosome reaction which is the reaction that occurs in the acrosome of the sperm as it approaches the egg.
Cortical reaction	The cortical reaction occurs in fertilisation when a sperm cell unites with the egg's plasma membrane, (zona reaction).This reaction leads to a modification of the zona pellucida that blocks polyspermy; enzymes released by cortical granules digest sperm receptor proteins ZP2 and ZP3 so that they can no longer bind sperm, in mammals.

Chapter 6. Fertilization

CHAPTER HIGHLIGHTS & NOTES: KEY TERMS, PEOPLE, PLACES, CONCEPTS

	The cortical reaction is exocytosis of the egg's cortical granules. Cortical granules are secretory vesicles that reside just below the egg's plasma membrane.
Mitosis	Mitosis is the process by which a eukaryotic cell separates the chromosomes in its cell nucleus into two identical sets in two nuclei. It is generally followed immediately by cytokinesis, which divides the nuclei, cytoplasm, organelles and cell membrane into two cells containing roughly equal shares of these cellular components. Mitosis and cytokinesis together define the mitotic (M) phase of the cell cycle--the division of the mother cell into two daughter cells, genetically identical to each other and to their parent cell.
Ubiquitin	Ubiquitin is a small regulatory protein that has been found in almost all tissues (ubiquitously) of eukaryotic organisms. Among other functions, it directs protein recycling. Ubiquitin binds to proteins and labels them for destruction.
Sperm	The term sperm is derived from the Greek word sperma and refers to the male reproductive cells. In the types of sexual reproduction known as anisogamy and oogamy, there is a marked difference in the size of the gametes with the smaller one being termed the 'male' or sperm cell. A uniflagellar sperm cell that is motile is referred to as a spermatozoon, whereas a non-motile sperm cell is referred to as a spermatium.
Hydatidiform mole	Molar pregnancy is an abnormal form of pregnancy, wherein a non-viable, fertilized egg implants in the uterus, and thereby converts normal pregnancy processes into pathological ones. It is characterized by the presence of a hydatidiform mole. Molar pregnancies are categorized into partial and complete moles.
Decidualization	Decidualization is a characteristic of the endometrium of the pregnant uterus. It is a response of maternal cells to progesterone. Decidualization may be used to describe any change due to progesterone.
Endometrium	The endometrium is the inner membrane of the mammalian uterus. The endometrium is the innermost glandular layer and functions as a lining for the uterus, preventing adhesions between the opposed walls of the myometrium, thereby maintaining the patency of the uterine cavity. During the menstrual cycle or estrous cycle, the endometrium grows to a thick, blood vessel-rich, glandular tissue layer.
Ectopic pregnancy	An ectopic pregnancy, is a complication of pregnancy in which the embryo implants outside the uterine cavity. With rare exceptions, ectopic pregnancies are not viable. Furthermore, they are dangerous for the parent, since internal haemorrhage is a life threatening complication.

Chapter 6. Fertilization

CHAPTER HIGHLIGHTS & NOTES: KEY TERMS, PEOPLE, PLACES, CONCEPTS

Fetus	A fetus (sometimes spelled foetus or fœtus) is a stage in the development of viviparous organisms. This stage lies between the embryonic stage and birth. The fetuses of most mammals are situated similarly to the homo sapiens fetus within their mothers.
Gonadotropin	Glycoprotein hormone Gonadotropins are protein hormones secreted by gonadotrope cells of the pituitary gland of vertebrates. This is a family of proteins, which include the mammalian hormones follitropin (FSH), lutropin (LH), placental chorionic gonadotropins hCG and eCG and chorionic gonadotropin as well as at least two forms of fish gonadotropins. These hormones are central to the complex endocrine system that regulates normal growth, sexual development, and reproductive function.
Immune system	The immune system is a system of biological structures and processes within an organism that protects against disease. To function properly, an immune system must detect a wide variety of agents, from viruses to parasitic worms, and distinguish them from the organism's own healthy tissue. Pathogens can rapidly evolve and adapt to avoid detection and neutralization by the immune system.
Fallopian tube	The Fallopian Tube, uterine tubes, and salpinges are two very fine tubes lined with ciliated epithelia, leading from the ovaries of female mammals into the uterus, via the utero-tubal junction. In non-mammalian vertebrates, the equivalent structures are the oviducts. In a woman's body the tube allows passage of the egg from the ovary to the uterus.
Pelvic inflammatory disease	Pelvic inflammatory disease is a generic term for inflammation of the uterus, fallopian tubes, and/or ovaries as it progresses to scar formation with adhesions to nearby tissues and organs. This may lead to infections. pelvic\ inflammatory\ disease is a vague term and can refer to viral, fungal, parasitic, though most often bacterial infections.
Polycystic ovary syndrome	Polycystic ovary syndrome is one of the most common female endocrine disorders. PCOS is a complex, heterogeneous disorder of uncertain etiology, but there is strong evidence that it can to a large degree be classified as a genetic disease. PCOS produces symptoms in approximately 5% to 10% of women of reproductive age (12-45 years old).

Chapter 6. Fertilization

CHAPTER HIGHLIGHTS & NOTES: KEY TERMS, PEOPLE, PLACES, CONCEPTS

Prostaglandin	A prostaglandin is any member of a group of lipid compounds that are derived enzymatically from fatty acids and have important functions in the animal body. Every prostaglandin contains 20 carbon atoms, including a 5-carbon ring.
	They are mediators and have a variety of strong physiological effects, such as regulating the contraction and relaxation of smooth muscle tissue. Prostaglandins are not hormones, but autocrine or paracrine, which are locally acting messenger molecules. They differ from hormones in that they are not produced at a discrete site but in many places throughout the human body. Also, their target cells are present in the immediate vicinity of the site of their secretion (of which there are many).
Female infertility	Female infertility refers to infertility in female humans.
	Causes or factors of female infertility can basically be classified regarding whether they are acquired or genetic, or strictly by location. Acquired versus genetic
	Although causes of female infertility can be classified as acquired versus genetic, female infertility is usually more or less a combination of nature and nurture.
Male infertility	Male infertility refers to the inability of a male to achieve a pregnancy in a fertile female. In humans it accounts for 40-50% of infertility. Male infertility is commonly due to deficiencies in the semen, and semen quality is used as a surrogate measure of male fecundity.
Platelet-activating factor	Platelet-activating factor, platelet\ activating\ factor-acether or AGEPC (acetyl-glyceryl-ether-phosphorylcholine) is a potent phospholipid activator and mediator of many leukocyte functions, including platelet aggregation and degranulation, inflammation, and anaphylaxis. It is also involved in changes to vascular permeability, the oxidative burst, chemotaxis of leukocytes, as well as augmentation of arachidonic acid metabolism in phagocytes.
	It is produced in response to specific stimuli by a variety of cell types, including neutrophils, basophils, injured tissue, monocytes/macrophages, platelets, and endothelial cells.
Assisted reproductive technology	Assisted reproductive technology is a general term referring to methods used to achieve pregnancy by artificial or partially artificial means. It is reproductive technology used primarily in infertility treatments. Some forms of ART are also used in fertile couples for genetic reasons.
Reproductive technology	Reproductive technology encompasses all current and anticipated uses of technology in human and animal reproduction, including assisted reproductive technology, contraception and others.
	Assisted reproductive technology is the use of reproductive technology to treat infertility.

Chapter 6. Fertilization

CHAPTER HIGHLIGHTS & NOTES: KEY TERMS, PEOPLE, PLACES, CONCEPTS

Angelman syndrome	Angelman syndrome is a neuro-genetic disorder characterized by intellectual and developmental disability, sleep disturbance, seizures, jerky movements (especially hand-flapping), frequent laughter or smiling, and usually a happy demeanor. AS is a classic example of genomic imprinting in that it is usually caused by deletion or inactivation of genes on the maternally inherited chromosome 15 while the paternal copy, which may be of normal sequence, is imprinted and therefore silenced. The sister syndrome, Prader-Willi syndrome, is caused by a similar loss of paternally inherited genes and maternal imprinting.
Ejaculatory duct	The ejaculatory ducts (ductus ejaculatorii) are paired structures in male anatomy. Each ejaculatory duct is formed by the union of the vas deferens with the duct of the seminal vesicle. They pass through the prostate, and open into the urethra at the Colliculus seminalis.
Percutaneous epididymal sperm aspiration	Percutaneous epididymal sperm aspiration is a technique used to determine sperm counts in the event of a possible blockage of the vas deferens. A small needle is inserted through the skin of the scrotum to collect sperm from the epididymis, where sperm are usually stored after production in the testes. It can also be used to extract sperm for intracytoplasmic sperm injection (ICSI).

CHAPTER QUIZ: KEY TERMS, PEOPLE, PLACES, CONCEPTS

1. A _____ is a cell that fuses with another cell during fertilization (conception) in organisms that reproduce sexually. In species that produce two morphologically distinct types of _____s, and in which each individual produces only one type, a female is any individual that produces the larger type of _____--called an ovum --and a male produces the smaller tadpole-like type--called a sperm. This is an example of anisogamy or heterogamy, the condition wherein females and males produce _____s of different sizes (this is the case in humans; the human ovum has approximately 100,000 times the volume of a single human sperm cell).

    a. Gonocyte
    b. Gynoecium
    c. Gamete
    d. Hydatid of Morgagni

2. . _____ is a condition in which a man ejaculates earlier than he or his partner would like him to. _____ is also known as rapid ejaculation, rapid climax, premature climax, or early ejaculation. Masters and Johnson defines PE as the condition in which a man ejaculates before his sex partner achieves orgasm, in more than fifty percent of their sexual encounters.

    a. Prostate-specific antigen

Chapter 6. Fertilization

CHAPTER QUIZ: KEY TERMS, PEOPLE, PLACES, CONCEPTS

    b. Sexual health clinic
    c. Premature ejaculation
    d. Spermatid

3. _____ is sexual dysfunction characterized by the inability to develop or maintain an erection of the penis during sexual performance.

   A penile erection is the hydraulic effect of blood entering and being retained in sponge-like bodies within the penis. The process is often initiated as a result of sexual arousal, when signals are transmitted from the brain to nerves in the penis.

       a. Izonsteride
       b. Alosetron
       c. Adhesion
       d. Erectile dysfunction

4. _____ is a generic term for inflammation of the uterus, fallopian tubes, and/or ovaries as it progresses to scar formation with adhesions to nearby tissues and organs. This may lead to infections. pelvic\ inflammatory\ disease is a vague term and can refer to viral, fungal, parasitic, though most often bacterial infections.

       a. Pelvic inflammatory disease
       b. Sexual health clinic
       c. Tuskegee syphilis experiment
       d. Vaginal microbicide

5. _____ ( ) is a mammalian hormone that acts primarily as a neuromodulator in the brain.

   _____ is best known for its roles in sexual reproduction, in particular during and after childbirth. It is released in large amounts after distension of the cervix and uterus during labor, facilitating birth, and after stimulation of the nipples, facilitating breastfeeding.

       a. Izonsteride
       b. Ejaculation
       c. Erotic sexual denial
       d. Oxytocin

ANSWER KEY
Chapter 6. Fertilization

1. c
2. c
3. d
4. a
5. d

You can take the complete Chapter Practice Test

for Chapter 6. Fertilization
on all key terms, persons, places, and concepts.

Online 99 Cents

http://www.epub5042.32.20612.6.cram101.com/

Use www.Cram101.com for all your study needs

including Cram101's online interactive problem solving labs in

chemistry, statistics, mathematics, and more.

Chapter 7. Overview of human genetics and genetic disorders

CHAPTER OUTLINE: KEY TERMS, PEOPLE, PLACES, CONCEPTS

	Genetics
	Eugenics
	Gene therapy
	Mitosis
	Amino acid
	Human genome
	Haemophilia
	Mutation
	Sickle-cell disease
	Y chromosome
	Allele
	Influenza
	Dominance
	Dwarfism
	Cystic fibrosis
	Blood flow
	Nervous system
	Angelman syndrome
	Alpha-fetoprotein

Chapter 7. Overview of human genetics and genetic disorders
CHAPTER OUTLINE: KEY TERMS, PEOPLE, PLACES, CONCEPTS

	Polymerase chain reaction
	Triple test
	Amniotic fluid
	Ectopic pregnancy
	Gonadotropin
	Amniocentesis
	Chorionic villus sampling

CHAPTER HIGHLIGHTS & NOTES: KEY TERMS, PEOPLE, PLACES, CONCEPTS

Genetics	Genetics, a discipline of biology, is the science of genes, heredity, and variation in living organisms.
	Genetics deals with the molecular structure and function of genes, with gene behavior in the context of a cell or organism (e.g. dominance and epigenetics), with patterns of inheritance from parent to offspring, and with gene distribution, variation and change in populations. Given that genes are universal to living organisms, genetics can be applied to the study of any living system from viruses and bacteria, through plants (especially crops) to humans (for example in Medical Genetics).
Eugenics	Eugenics is the 'applied science or the bio-social movement which advocates the use of practices aimed at improving the genetic composition of a population', usually referring to the manipulation of human populations. The origins of the concept of eugenics began with certain interpretations of Mendelian inheritance, and the theories of August Weismann. Historically, many of the practitioners of eugenics viewed eugenics as a science, not necessarily restricted to human populations; this embraced the views of Darwinism and Social Darwinism.
Gene therapy	Gene therapy is the use of DNA as a pharmaceutical agent to treat disease.

Chapter 7. Overview of human genetics and genetic disorders

CHAPTER HIGHLIGHTS & NOTES: KEY TERMS, PEOPLE, PLACES, CONCEPTS

	It derives its name from the idea that DNA can be used to supplement or alter genes within an individual's cells as a therapy to treat disease. The most common form of gene therapy involves using DNA that encodes a functional, therapeutic gene in order to replace a mutated gene.
Mitosis	Mitosis is the process by which a eukaryotic cell separates the chromosomes in its cell nucleus into two identical sets in two nuclei. It is generally followed immediately by cytokinesis, which divides the nuclei, cytoplasm, organelles and cell membrane into two cells containing roughly equal shares of these cellular components. Mitosis and cytokinesis together define the mitotic (M) phase of the cell cycle--the division of the mother cell into two daughter cells, genetically identical to each other and to their parent cell.
Amino acid	Amino acids are molecules containing an amine group, a carboxylic acid group and a side-chain that varies between different amino acids. The key elements of an amino acid are carbon, hydrogen, oxygen, and nitrogen. They are particularly important in biochemistry, where the term usually refers to alpha-amino acids.
Human genome	The human genome is the genome of Homo sapiens, which is stored on 23 chromosome pairs. 22 of these are autosomal chromosome pairs, while the remaining pair is sex-determining. The haploid human genome occupies a total of just over 3 billion DNA base pairs. The Human Genome Project (HGP) produced a reference sequence of the euchromatic human genome, which is used worldwide in biomedical sciences. The haploid human genome contains ca. 23,000 protein-coding genes, far fewer than had been expected before its sequencing. In fact, only about 1.5% of the genome codes for proteins, while the rest consists of non-coding RNA genes, regulatory sequences, introns, and noncoding DNA (once known as 'junk DNA').
Haemophilia	Haemophilia is a group of hereditary genetic disorders that impair the body's ability to control blood clotting or coagulation, which is used to stop bleeding when a blood vessel is broken. Haemophilia A (clotting factor VIII deficiency) is the most common form of the disorder, occurring at about 1 in 5,000-10,000 male births. Haemophilia B (factor IX deficiency) occurs at about 1 in about 20,000-34,000 male births.
Mutation	In molecular biology and genetics, mutations are changes in a genomic sequence: the DNA sequence of a cell's genome or the DNA or RNA sequence of a virus. They can be defined as sudden and spontaneous changes in the cell. Mutations are caused by radiation, viruses, transposons and mutagenic chemicals, as well as errors that occur during meiosis or DNA replication. They can also be induced by the organism itself, by cellular processes such as hypermutation.

Chapter 7. Overview of human genetics and genetic disorders

CHAPTER HIGHLIGHTS & NOTES: KEY TERMS, PEOPLE, PLACES, CONCEPTS

Sickle-cell disease	Sickle-cell disease is an autosomal recessive genetic blood disorder, with overdominance, characterized by red blood cells that assume an abnormal, rigid, sickle shape. Sickling decreases the cells' flexibility and results in a risk of various complications. The sickling occurs because of a mutation in the haemoglobin gene. Life expectancy is shortened, with studies reporting an average life expectancy of 42 in males and 48 in females.
Y chromosome	The Y chromosome is one of the two sex-determining chromosomes in most mammals, including humans. In mammals, it contains the gene SRY, which triggers testis development if present. The human Y chromosome is composed of about 50 million base pairs.
Allele	An allele is one of two or more forms of a gene. Sometimes, different alleles can result in different traits, such as color. Other times, different alleles will have the same result in the expression of a gene. Most multicellular organisms have two sets of chromosomes, that is, they are diploid. These chromosomes are referred to as homologous chromosomes. Diploid organisms have one copy of each gene (and one allele) on each chromosome. If both alleles are the same, they are homozygotes. If the alleles are different, they are heterozygotes.
Influenza	Influenza, commonly referred to as the flu, is an infectious disease caused by RNA viruses of the family Orthomyxoviridae (the influenza viruses), that affects birds and mammals. The most common symptoms of the disease are chills, fever, sore throat, muscle pains, severe headache, coughing, weakness/fatigue and general discomfort. Although it is often confused with other influenza-like illnesses, especially the common cold, influenza is a more severe disease than the common cold and is caused by a different type of virus.
Dominance	Dominance in genetics is a relationship between two variant forms (alleles) of a single gene, in which one allele masks the expression of the other in influencing some trait. In the simplest case, if a gene exists in two allelic forms (A & B), three combinations of alleles (genotypes) are possible: AA, AB, and BB. If AB individuals (heterozygotes) show the same form of the trait (phenotype) as AA individuals (homozygotes), and BB homozygotes show an alternative phenotype, allele A is said to dominate or be dominant to allele B, and B is said to be recessive to A.
Dwarfism	Dwarfism is when an individual is short in stature resulting from a medical condition caused by problems that arise in the pituitary gland in which the growth of the individual is very slowed or delayed. Dwarfism is sometimes defined as an adult height of less than 147 cm (58 inches), although this definition is problematic because short stature in itself is not a disorder. For example, in pygmy populations, a body height of less than 150 cm (59 inches) is normal.

Chapter 7. Overview of human genetics and genetic disorders

CHAPTER HIGHLIGHTS & NOTES: KEY TERMS, PEOPLE, PLACES, CONCEPTS

Cystic fibrosis	Cystic fibrosis is a recessive genetic disease affecting most critically the lungs, and also the pancreas, liver, and intestine. It is characterized by abnormal transport of chloride and sodium across epithelium, leading to thick, viscous secretions. The name cystic fibrosis refers to the characteristic scarring (fibrosis) and cyst formation within the pancreas, first recognized in the 1930s.
Blood flow	Blood flow is the continuous running of blood in the cardiovascular system. The human body is made up of several processes all carrying out various functions. We have the gastrointestinal system which aids the digestion and the absorption of food.
Nervous system	The nervous system is an organ system containing a network of specialized cells called neurons that coordinate the actions of an animal and transmit signals between different parts of its body. In most animals the nervous system consists of two parts, central and peripheral. The central nervous system of vertebrates (such as humans) contains the brain, spinal cord, and retina. The peripheral nervous system consists of sensory neurons, clusters of neurons called ganglia, and nerves connecting them to each other and to the central nervous system. These regions are all interconnected by means of complex neural pathways. The enteric nervous system, a subsystem of the peripheral nervous system, has the capacity, even when severed from the rest of the nervous system through its primary connection by the vagus nerve, to function independently in controlling the gastrointestinal system.
Angelman syndrome	Angelman syndrome is a neuro-genetic disorder characterized by intellectual and developmental disability, sleep disturbance, seizures, jerky movements (especially hand-flapping), frequent laughter or smiling, and usually a happy demeanor. AS is a classic example of genomic imprinting in that it is usually caused by deletion or inactivation of genes on the maternally inherited chromosome 15 while the paternal copy, which may be of normal sequence, is imprinted and therefore silenced. The sister syndrome, Prader-Willi syndrome, is caused by a similar loss of paternally inherited genes and maternal imprinting.
Alpha-fetoprotein	Alpha-fetoprotein is a protein that in humans is encoded by the AFP gene. AFP is a major plasma protein produced by the yolk sac and the liver during fetal development that is thought to be the fetal form of serum albumin. The AFP gene is located on the q arm of chromosome 4 (4q25).
Polymerase chain reaction	The polymerase chain reaction is a scientific technique in molecular biology to amplify a single or a few copies of a piece of DNA across several orders of magnitude, generating thousands to millions of copies of a particular DNA sequence.

Chapter 7. Overview of human genetics and genetic disorders

CHAPTER HIGHLIGHTS & NOTES: KEY TERMS, PEOPLE, PLACES, CONCEPTS

	Developed in 1983 by Kary Mullis, PCR is now a common and often indispensable technique used in medical and biological research labs for a variety of applications. These include DNA cloning for sequencing, DNA-based phylogeny, or functional analysis of genes; the diagnosis of hereditary diseases; the identification of genetic fingerprints (used in forensic sciences and paternity testing); and the detection and diagnosis of infectious diseases.
Triple test	The triple test, the Kettering test or the Bart's test, is an investigation performed during pregnancy in the second trimester to classify a patient as either high-risk or low-risk for chromosomal abnormalities (and neural tube defects).
	The term 'multiple-marker screening test' is sometimes used instead. This term can encompass the 'double test' and 'quadruple test' (described below).
Amniotic fluid	Amniotic fluid is the nourishing and protecting liquid contained by the amniotic sac of a pregnant woman.
	From the very beginning of the formation of the extracoelomal cavity, amniotic fluid [AF] can be detected. This firstly water-like fluid originates from the maternal plasma, and passes through the fetal membranes by osmotic and hydrostatic forces.
Ectopic pregnancy	An ectopic pregnancy, is a complication of pregnancy in which the embryo implants outside the uterine cavity. With rare exceptions, ectopic pregnancies are not viable. Furthermore, they are dangerous for the parent, since internal haemorrhage is a life threatening complication.
Gonadotropin	Glycoprotein hormone
	Gonadotropins are protein hormones secreted by gonadotrope cells of the pituitary gland of vertebrates. This is a family of proteins, which include the mammalian hormones follitropin (FSH), lutropin (LH), placental chorionic gonadotropins hCG and eCG and chorionic gonadotropin as well as at least two forms of fish gonadotropins. These hormones are central to the complex endocrine system that regulates normal growth, sexual development, and reproductive function.
Amniocentesis	Amniocentesis is a medical procedure used in prenatal diagnosis of chromosomal abnormalities and fetal infections, in which a small amount of amniotic fluid, which contains fetal tissues, is sampled from the amnion or amniotic sac surrounding a developing fetus, and the fetal DNA is examined for genetic abnormalities. Using this process the sex of a child can also be determined and hence this procedure has some legal restrictions in some gender biased countries.

Chapter 7. Overview of human genetics and genetic disorders

CHAPTER HIGHLIGHTS & NOTES: KEY TERMS, PEOPLE, PLACES, CONCEPTS

Chorionic villus sampling	Chorionic villus sampling sometimes misspelled 'chorionic villous sampling', is a form of prenatal diagnosis to determine chromosomal or genetic disorders in the fetus. It entails sampling of the chorionic villus (placental tissue) and testing it for chromosomal abnormalities, usually with FISH or PCR. chorionic\ villus\ sampling usually takes place at 10-12 weeks' gestation, earlier than amniocentesis (14-16 weeks). It is the preferred technique before 15 weeks.

CHAPTER QUIZ: KEY TERMS, PEOPLE, PLACES, CONCEPTS

1. An _____ is one of two or more forms of a gene. Sometimes, different _____s can result in different traits, such as color. Other times, different _____s will have the same result in the expression of a gene.

    Most multicellular organisms have two sets of chromosomes, that is, they are diploid. These chromosomes are referred to as homologous chromosomes. Diploid organisms have one copy of each gene (and one _____) on each chromosome. If both _____s are the same, they are homozygotes. If the _____s are different, they are heterozygotes.

    a. Izonsteride
    b. Alosetron
    c. Allele
    d. Hayden Act

2. _____, a discipline of biology, is the science of genes, heredity, and variation in living organisms.

    _____ deals with the molecular structure and function of genes, with gene behavior in the context of a cell or organism (e.g. dominance and epigenetics), with patterns of inheritance from parent to offspring, and with gene distribution, variation and change in populations. Given that genes are universal to living organisms, _____ can be applied to the study of any living system from viruses and bacteria, through plants (especially crops) to humans (for example in Medical _____).

    a. Lysis
    b. Homeostasis
    c. Genetics
    d. magnetic resonance imaging

3. . The _____ is a scientific technique in molecular biology to amplify a single or a few copies of a piece of DNA across several orders of magnitude, generating thousands to millions of copies of a particular DNA sequence.

Chapter 7. Overview of human genetics and genetic disorders

CHAPTER QUIZ: KEY TERMS, PEOPLE, PLACES, CONCEPTS

Developed in 1983 by Kary Mullis, PCR is now a common and often indispensable technique used in medical and biological research labs for a variety of applications. These include DNA cloning for sequencing, DNA-based phylogeny, or functional analysis of genes; the diagnosis of hereditary diseases; the identification of genetic fingerprints (used in forensic sciences and paternity testing); and the detection and diagnosis of infectious diseases.

a. Primer dimer
b. Procell
c. Progenitor cell
d. Polymerase chain reaction

4. _____s are molecules containing an amine group, a carboxylic acid group and a side-chain that varies between different _____s. The key elements of an _____ are carbon, hydrogen, oxygen, and nitrogen. They are particularly important in biochemistry, where the term usually refers to alpha-_____s.

a. Izonsteride
b. Amino acid
c. Hastings Center Report
d. Hayden Act

5. _____ is the 'applied science or the bio-social movement which advocates the use of practices aimed at improving the genetic composition of a population', usually referring to the manipulation of human populations. The origins of the concept of _____ began with certain interpretations of Mendelian inheritance, and the theories of August Weismann. Historically, many of the practitioners of _____ viewed _____ as a science, not necessarily restricted to human populations; this embraced the views of Darwinism and Social Darwinism.

a. Euthenics
b. Eugenics
c. Utilitarian bioethics
d. magnetic resonance imaging

ANSWER KEY
Chapter 7. Overview of human genetics and genetic disorders

1. c
2. c
3. d
4. b
5. b

You can take the complete Chapter Practice Test

for Chapter 7. Overview of human genetics and genetic disorders
on all key terms, persons, places, and concepts.

Online 99 Cents

http://www.epub5042.32.20612.7.cram101.com/

Use www.Cram101.com for all your study needs

including Cram101's online interactive problem solving labs in

chemistry, statistics, mathematics, and more.

Chapter 8. The placenta

CHAPTER OUTLINE: KEY TERMS, PEOPLE, PLACES, CONCEPTS

	Placenta
	Pre-eclampsia
	Blood flow
	Cardiac output
	Blood cell
	Amniotic fluid
	Oligohydramnios
	Polyhydramnios
	Alpha-fetoprotein
	Angiogenesis
	Vasculogenesis
	Small for gestational age
	Gestational age
	Nitric oxide
	Amino acid
	Fatty acid
	Steroid hormone
	Ectopic pregnancy
	Gonadotropin

Chapter 8. The placenta
CHAPTER OUTLINE: KEY TERMS, PEOPLE, PLACES, CONCEPTS

- Secretion
- Allantois
- Human placental lactogen
- Yolk sac
- Growth hormone
- Placental lactogen
- Diabetes mellitus
- Umbilical artery
- Circumvallate placenta

CHAPTER HIGHLIGHTS & NOTES: KEY TERMS, PEOPLE, PLACES, CONCEPTS

Placenta	The placenta is an organ that connects the developing fetus to the uterine wall to allow nutrient uptake, waste elimination, and gas exchange via the mother's blood supply. 'True' placentas are a defining characteristic of eutherian or 'placental' mammals, but are also found in some snakes and lizards with varying levels of development up to mammalian levels. Note, however, that the homology of such structures in various viviparous organisms is debatable at best and, in invertebrates such as Arthropoda, is definitely analogous at best.
Pre-eclampsia	Pre-eclampsia is a medical condition in which hypertension arises in pregnancy (pregnancy-induced hypertension) in association with significant amounts of protein in the urine. Pre-eclampsia refers to a set of symptoms rather than any causative factor, and there are many different causes for the condition. It appears likely that there are substances from the placenta that can cause endothelial dysfunction in the maternal blood vessels of susceptible women.

Chapter 8. The placenta

CHAPTER HIGHLIGHTS & NOTES: KEY TERMS, PEOPLE, PLACES, CONCEPTS

Blood flow	Blood flow is the continuous running of blood in the cardiovascular system. The human body is made up of several processes all carrying out various functions. We have the gastrointestinal system which aids the digestion and the absorption of food.
Cardiac output	Cardiac output is the volume of blood being pumped by the heart, in particular by a left or right ventricle in the time interval of one minute. CO may be measured in many ways, for example dm^3/min (1 dm^3 equals 1000 cm^3 or 1 litre). Q is furthermore the combined sum of output from the right ventricle and the output from the left ventricle during the phase of systole of the heart.
Blood cell	A blood cell, is a cell produced by haematopoiesis and normally found in blood. In mammals, these fall into three general categories:•Red blood cells -- Erythrocytes•White blood cells -- Leukocytes•Platelets -- Thrombocytes. Together, these three kinds of blood cells add up to a total 45% of the blood tissue by volume, with the remaining 55% of the volume composed of plasma, the liquid component of blood. This volume percentage (e.g., 45%) of cells to total volume is called hematocrit, determined by centrifuge or flow cytometry.
Amniotic fluid	Amniotic fluid is the nourishing and protecting liquid contained by the amniotic sac of a pregnant woman. From the very beginning of the formation of the extracoelomal cavity, amniotic fluid [AF] can be detected. This firstly water-like fluid originates from the maternal plasma, and passes through the fetal membranes by osmotic and hydrostatic forces.
Oligohydramnios	Oligohydramnios is a condition in pregnancy characterized by a deficiency of amniotic fluid. It is the opposite of polyhydramnios. The common clinical features are smaller symphysio fundal height, fetal malpresentation, undue prominence fetal parts and reduced amount of amniotic fluid.
Polyhydramnios	Polyhydramnios is a medical condition describing an excess of amniotic fluid in the amniotic sac. It is seen in about 1% of pregnancies,. It is typically diagnosed when the amniotic fluid index (AFI) is greater than 24 cm There are two clinical varieties of polyhydramnios:•Chronic polyhydramnios where excess amniotic fluid accumulates gradually.•Acute polyhydramnios where excess amniotic fluid collects rapidly. The opposite to polyhydramnios is oligohydramnios, a deficiency in amniotic fluid.
Alpha-fetoprotein	Alpha-fetoprotein is a protein that in humans is encoded by the AFP gene.

Chapter 8. The placenta

CHAPTER HIGHLIGHTS & NOTES: KEY TERMS, PEOPLE, PLACES, CONCEPTS

	AFP is a major plasma protein produced by the yolk sac and the liver during fetal development that is thought to be the fetal form of serum albumin. The AFP gene is located on the q arm of chromosome 4 (4q25).
Angiogenesis	Angiogenesis is the physiological process involving the growth of new blood vessels from pre-existing vessels. Though there has been some debate over terminology, vasculogenesis is the term used for spontaneous blood-vessel formation, and intussusception is the term for the formation of new blood vessels by the splitting of existing ones. Angiogenesis is a normal and vital process in growth and development, as well as in wound healing and in granulation tissue.
Vasculogenesis	Vasculogenesis is the process of blood vessel formation occurring by a de novo production of endothelial cells. Though similar to angiogenesis, the two are different in one aspect: The term angiogenesis denotes the formation of new blood vessels from pre-existing ones, whereas vasculogenesis is the term used for the formation of new blood vessels when there are no pre-existing ones. For example, if a monolayer of endothelial cells begins sprouting to form capillaries, angiogenesis is occurring.
Small for gestational age	Small for gestational age babies are those who are smaller in size than normal for the baby's sex and gestational age, most commonly defined as a weight below the 10th percentile for the gestational age. Not all fetuses that are SGA are pathologically growth restricted and, in fact, may be constitutionally small. If small for gestational age babies have been the subject of intrauterine growth restriction (IUGR), formerly known as intrauterine growth retardation, the term SGA associated with IUGR is used.
Gestational age	Gestational age is the age of an embryo or fetus . In humans, a common method of calculating gestational age starts counting either from the first day of the woman's last menstrual period (LMP) or from 14 days before conception (fertilization). Counting from the first day of the LMP involves the assumption that conception occurred 14 days later.
Nitric oxide	Nitric oxide is a chemical compound with chemical formula Nitric oxide. This diatomic gas is an important cell signaling molecule in mammals, including humans. It is also an extremely important intermediate in the chemical industry. Nitric oxide is an air pollutant produced by combustion of substances in air, like in automobile engines and fossil fuel power plants.

Chapter 8. The placenta

CHAPTER HIGHLIGHTS & NOTES: KEY TERMS, PEOPLE, PLACES, CONCEPTS

Amino acid	Amino acids are molecules containing an amine group, a carboxylic acid group and a side-chain that varies between different amino acids. The key elements of an amino acid are carbon, hydrogen, oxygen, and nitrogen. They are particularly important in biochemistry, where the term usually refers to alpha-amino acids.
Fatty acid	In chemistry, especially biochemistry, a fatty acid is a carboxylic acid with a long aliphatic tail (chain), which is either saturated or unsaturated. Most naturally occurring fatty acids have a chain of an even number of carbon atoms, from 4 to 28. Fatty acids are usually derived from triglycerides or phospholipids. When they are not attached to other molecules, they are known as 'free' fatty acids.
Steroid hormone	A steroid hormone is a steroid that acts as a hormone. Steroid hormones can be grouped into five groups by the receptors to which they bind: glucocorticoids, mineralocorticoids, androgens, estrogens, and progestogens. Vitamin D derivatives are a sixth closely related hormone system with homologous receptors, though technically sterols rather than steroids.
Ectopic pregnancy	An ectopic pregnancy, is a complication of pregnancy in which the embryo implants outside the uterine cavity. With rare exceptions, ectopic pregnancies are not viable. Furthermore, they are dangerous for the parent, since internal haemorrhage is a life threatening complication.
Gonadotropin	Glycoprotein hormone Gonadotropins are protein hormones secreted by gonadotrope cells of the pituitary gland of vertebrates. This is a family of proteins, which include the mammalian hormones follitropin (FSH), lutropin (LH), placental chorionic gonadotropins hCG and eCG and chorionic gonadotropin as well as at least two forms of fish gonadotropins. These hormones are central to the complex endocrine system that regulates normal growth, sexual development, and reproductive function.
Secretion	Secretion is the process of elaborating, releasing, and oozing chemicals, or a secreted chemical substance from a cell or gland. In contrast to excretion, the substance may have a certain function, rather than being a waste product. Many cells contain this such as glucoma cells.
Allantois	Allantois is a part of a developing animal conceptus (which consists of all embryonic and extra-embryonic tissues). It helps the embryo exchange gases and handle liquid waste. The allantois, along with the amnion and chorion (other embryonic membranes), identify humans, and other mammals, as amniotes.

Chapter 8. The placenta

CHAPTER HIGHLIGHTS & NOTES: KEY TERMS, PEOPLE, PLACES, CONCEPTS

Human placental lactogen	Human placental lactogen also called human chorionic somatomammotropin, is a polypeptide placental hormone. Its structure and function is similar to that of human growth hormone. It modifies the metabolic state of the mother during pregnancy to facilitate the energy supply of the fetus.
Yolk sac	The yolk sac is a membranous sac attached to an embryo, providing early nourishment in the form of yolk in bony fishes, sharks, reptiles, birds, and primitive mammals. It functions as the developmental circulatory system of the human embryo, before internal circulation begins. It is the first element seen in the gestational sac during pregnancy, usually at 5 weeks gestation.
Growth hormone	Growth hormone is a peptide hormone that stimulates growth, cell reproduction and regeneration in humans and other animals. Growth hormone is a 191-amino acid, single-chain polypeptide that is synthesized, stored, and secreted by the somatotroph cells within the lateral wings of the anterior pituitary gland. Somatotropin (STH) refers to the growth hormone 1 produced naturally in animals, whereas the term somatropin refers to growth hormone produced by recombinant DNA technology, and is abbreviated 'HGH' in humans.
Placental lactogen	Placental lactogen is a polypeptide placental hormone. Its structure and function is similar to that of growth hormone. It modifies the metabolic state of the mother during pregnancy to facilitate the energy supply of the fetus.
Diabetes mellitus	Diabetes mellitus, often simply referred to as diabetes, is a group of metabolic diseases in which a person has high blood sugar, either because the body does not produce enough insulin, or because cells do not respond to the insulin that is produced. This high blood sugar produces the classical symptoms of polyuria (frequent urination), polydipsia (increased thirst) and polyphagia (increased hunger). The three main types of diabetes mellitus are:•Type 1 DM results from the body's failure to produce insulin, and presently requires the person to inject insulin.
Umbilical artery	The umbilical artery is a paired artery (with one for each half of the body) that is found in the abdominal and pelvic regions. In the fetus, it extends into the umbilical cord. Umbilical arteries supply deoxygenated blood from the fetus to the placenta in the umbilical cord.
Circumvallate placenta	Circumvallate placenta is a placental disease in which the fetal membranes (chorion and amnion) 'double back' on the fetal side around the edge of the placenta. After delivery, a circumvallate placenta has a thick ring of membranes on its fetal surface.

Chapter 8. The placenta

CHAPTER QUIZ: KEY TERMS, PEOPLE, PLACES, CONCEPTS

1. _____s are molecules containing an amine group, a carboxylic acid group and a side-chain that varies between different _____s. The key elements of an _____ are carbon, hydrogen, oxygen, and nitrogen. They are particularly important in biochemistry, where the term usually refers to alpha-_____s.

   a. Izonsteride
   b. Hemolytic disease of the newborn
   c. Home birth
   d. Amino acid

2. _____ is a medical condition in which hypertension arises in pregnancy (pregnancy-induced hypertension) in association with significant amounts of protein in the urine.

   _____ refers to a set of symptoms rather than any causative factor, and there are many different causes for the condition. It appears likely that there are substances from the placenta that can cause endothelial dysfunction in the maternal blood vessels of susceptible women.

   a. Pregnancy category
   b. Pre-eclampsia
   c. Prenatal nutrition
   d. Reproductive immunology

3. _____ is the process of elaborating, releasing, and oozing chemicals, or a secreted chemical substance from a cell or gland. In contrast to excretion, the substance may have a certain function, rather than being a waste product. Many cells contain this such as glucoma cells.

   a. Microbiology
   b. N-Acyl homoserine lactone
   c. Bacillus anthracis phage AP50
   d. Secretion

4. _____ is a peptide hormone that stimulates growth, cell reproduction and regeneration in humans and other animals. _____ is a 191-amino acid, single-chain polypeptide that is synthesized, stored, and secreted by the somatotroph cells within the lateral wings of the anterior pituitary gland. Somatotropin (STH) refers to the _____ 1 produced naturally in animals, whereas the term somatropin refers to _____ produced by recombinant DNA technology, and is abbreviated 'HGH' in humans.

   a. Luteinizing hormone
   b. Melanocyte-stimulating hormone
   c. Prolactin
   d. Growth hormone

5. . _____ is the physiological process involving the growth of new blood vessels from pre-existing vessels.

Chapter 8. The placenta

CHAPTER QUIZ: KEY TERMS, PEOPLE, PLACES, CONCEPTS

Though there has been some debate over terminology, vasculogenesis is the term used for spontaneous blood-vessel formation, and intussusception is the term for the formation of new blood vessels by the splitting of existing ones.

_____ is a normal and vital process in growth and development, as well as in wound healing and in granulation tissue.

a. Angiogenesis
b. Arteriogenesis
c. Estrogen receptor test
d. Puerperal fever

ANSWER KEY
Chapter 8. The placenta

1. d
2. b
3. d
4. d
5. a

You can take the complete Chapter Practice Test

for Chapter 8. The placenta
on all key terms, persons, places, and concepts.

Online 99 Cents

http://www.epub5042.32.20612.8.cram101.com/

Use www.Cram101.com for all your study needs

including Cram101's online interactive problem solving labs in

chemistry, statistics, mathematics, and more.

Chapter 9. Embryo development and fetal growth

CHAPTER OUTLINE: KEY TERMS, PEOPLE, PLACES, CONCEPTS

	Ejaculatory duct
	Yolk sac
	Cloacal membrane
	Conjoined twins
	Transcription factor
	Apoptosis
	Germ cell
	Nervous system
	Neural tube defect
	Fetus
	Angiogenesis
	Atrial septal defect
	Diaphragmatic hernia
	Heart development
	Lymphocyte
	Respiratory system
	Vasculogenesis
	Ventricular septal defect
	Capillary

Chapter 9. Embryo development and fetal growth
CHAPTER OUTLINE: KEY TERMS, PEOPLE, PLACES, CONCEPTS

	Kidney development
	Urinary system
	Renal agenesis
	Amniotic fluid
	Small for gestational age
	Gestational age
	Growth hormone
	Growth factor
	Glucocorticoid
	Maternal nutrition
	Infection
	Obesity
	Pre-eclampsia
	Immune system

Chapter 9. Embryo development and fetal growth

CHAPTER HIGHLIGHTS & NOTES: KEY TERMS, PEOPLE, PLACES, CONCEPTS

Ejaculatory duct	The ejaculatory ducts (ductus ejaculatorii) are paired structures in male anatomy. Each ejaculatory duct is formed by the union of the vas deferens with the duct of the seminal vesicle. They pass through the prostate, and open into the urethra at the Colliculus seminalis.
Yolk sac	The yolk sac is a membranous sac attached to an embryo, providing early nourishment in the form of yolk in bony fishes, sharks, reptiles, birds, and primitive mammals. It functions as the developmental circulatory system of the human embryo, before internal circulation begins. It is the first element seen in the gestational sac during pregnancy, usually at 5 weeks gestation.
Cloacal membrane	The cloacal membrane is the membrane that covers the embryonic cloaca when still in the development of the urinary and reproductive organs. It is formed by ectoderm and endoderm coming into contact with each other. After separation of the cloaca into the urogenital and anal parts, the cloacal membrane, in turn, is separated into a urogenital membrane and an anal membrane.
Conjoined twins	Conjoined twins are identical twins whose bodies are joined in utero. A rare phenomenon, the occurrence is estimated to range from 1 in 50,000 births to 1 in 100,000 births, with a somewhat higher incidence in Southwest Asia and Africa. Approximately half are stillborn, and a smaller fraction of pairs born alive have abnormalities incompatible with life. The overall survival rate for conjoined twins is approximately 25%.
Transcription factor	In molecular biology and genetics, a transcription factor is a protein that binds to specific DNA sequences, thereby controlling the flow (or transcription) of genetic information from DNA to mRNA. Transcription factors perform this function alone or with other proteins in a complex, by promoting (as an activator), or blocking (as a repressor) the recruitment of RNA polymerase (the enzyme that performs the transcription of genetic information from DNA to RNA) to specific genes. A defining feature of transcription factors is that they contain one or more DNA-binding domains (DBDs), which attach to specific sequences of DNA adjacent to the genes that they regulate. Additional proteins such as coactivators, chromatin remodelers, histone acetylases, deacetylases, kinases, and methylases, while also playing crucial roles in gene regulation, lack DNA-binding domains, and, therefore, are not classified as transcription factors.
Apoptosis	Apoptosis is the process of programmed cell death (PCD) that may occur in multicellular organisms. Biochemical events lead to characteristic cell changes (morphology) and death. These changes include blebbing, cell shrinkage, nuclear fragmentation, chromatin condensation, and chromosomal DNA fragmentation.

Chapter 9. Embryo development and fetal growth

CHAPTER HIGHLIGHTS & NOTES: KEY TERMS, PEOPLE, PLACES, CONCEPTS

Germ cell	A germ cell is any biological cell that gives rise to the gametes of an organism that reproduces sexually. In many animals, the germ cells originate near the gut of an embryo and migrate to the developing gonads. There, they undergo cell division of two types, mitosis and meiosis, followed by cellular differentiation into mature gametes, either eggs or sperm.
Nervous system	The nervous system is an organ system containing a network of specialized cells called neurons that coordinate the actions of an animal and transmit signals between different parts of its body. In most animals the nervous system consists of two parts, central and peripheral. The central nervous system of vertebrates (such as humans) contains the brain, spinal cord, and retina. The peripheral nervous system consists of sensory neurons, clusters of neurons called ganglia, and nerves connecting them to each other and to the central nervous system. These regions are all interconnected by means of complex neural pathways. The enteric nervous system, a subsystem of the peripheral nervous system, has the capacity, even when severed from the rest of the nervous system through its primary connection by the vagus nerve, to function independently in controlling the gastrointestinal system.
Neural tube defect	Neural tube defects are one of the most common birth defects, occurring in approximately one in 1,000 live births in the United States. An Neural tube defect is an opening in the spinal cord or brain that occurs very early in human development. In about the 3rd or 4th weeks of pregnancy, specialized cells on the dorsal side of the fetus begin to fuse and form the neural tube.
Fetus	A fetus (sometimes spelled foetus or fœtus) is a stage in the development of viviparous organisms. This stage lies between the embryonic stage and birth. The fetuses of most mammals are situated similarly to the homo sapiens fetus within their mothers.
Angiogenesis	Angiogenesis is the physiological process involving the growth of new blood vessels from pre-existing vessels. Though there has been some debate over terminology, vasculogenesis is the term used for spontaneous blood-vessel formation, and intussusception is the term for the formation of new blood vessels by the splitting of existing ones. Angiogenesis is a normal and vital process in growth and development, as well as in wound healing and in granulation tissue.
Atrial septal defect	Atrial septal defect is a form of congenital heart defect that enables blood flow between the left and right atria via the interatrial septum. The interatrial septum is the tissue that divides the right and left atria. Without this septum, or if there is a defect in this septum, it is possible for blood to travel from the left side of the heart to the right side of the heart, or vice versa.

Chapter 9. Embryo development and fetal growth

CHAPTER HIGHLIGHTS & NOTES: KEY TERMS, PEOPLE, PLACES, CONCEPTS

Diaphragmatic hernia	Diaphragmatic hernia is a defect or hole in the diaphragm that allows the abdominal contents to move into the chest cavity. Treatment is usually surgical.
	The following types of diaphragmatic hernia exist:•Congenital diaphragmatic hernia •Morgagni's hernia•Bochdalek hernia•Hiatal hernia•Iatrogenic diaphragmatic hernia•Traumatic diaphragmatic hernia
	Signs and symptoms
	A scaphoid abdomen (sucked inwards) may be the presenting symptom in a newborn.
Heart development	The heart is the first functional organ in a vertebrate embryo. There are 5 stages to heart development.
	The lateral plate mesoderm delaminates to form two layers: the dorsal somatic (parietal) mesoderm and the ventral splanchnic (visceral) mesoderm.
Lymphocyte	A lymphocyte is a type of white blood cell in the vertebrate immune system.
	Under the microscope, lymphocytes can be divided into large granular lymphocytes and small lymphocytes. Large granular lymphocytes include natural killer cells (NK cells). Small lymphocytes consist of T cells and B cells.
Respiratory system	The respiratory system is the biological system of an organism that introduces respiratory gases to the interior and performs gas exchange. In humans and other mammals, the anatomical features of the respiratory system include airways, lungs, and the respiratory muscles. Molecules of oxygen and carbon dioxide are passively exchanged, by diffusion, between the gaseous external environment and the blood.
Vasculogenesis	Vasculogenesis is the process of blood vessel formation occurring by a de novo production of endothelial cells.
	Though similar to angiogenesis, the two are different in one aspect: The term angiogenesis denotes the formation of new blood vessels from pre-existing ones, whereas vasculogenesis is the term used for the formation of new blood vessels when there are no pre-existing ones. For example, if a monolayer of endothelial cells begins sprouting to form capillaries, angiogenesis is occurring.
Ventricular septal defect	A ventricular septal defect is a defect in the ventricular septum, the wall dividing the left and right ventricles of the heart.

Chapter 9. Embryo development and fetal growth

CHAPTER HIGHLIGHTS & NOTES: KEY TERMS, PEOPLE, PLACES, CONCEPTS

	The ventricular septum consists of an inferior muscular and superior membranous portion and is extensively innervated with conducting cardiomyocytes.
	The membranous portion, which is close to the atrioventricular node, is most commonly affected in adults and older children in the United States.
Capillary	Capillaries ( ) are the smallest of a body's blood vessels and are parts of the microcirculation. They are only 1 cell thick. These microvessels, measuring 5-10 μm in diameter, connect arterioles and venules, and enable the exchange of water, oxygen, carbon dioxide, and many other nutrients and waste chemical substances between blood and surrounding tissues. During embryological development, new capillaries are formed by vasculogenesis, the process of blood vessel formation occurring by a de novo production of endothelial cells and their formation into vascular tubes. The term angiogenesis denotes the formation of new capillaries from pre-existing blood vessels.
	Blood flows away from the heart via arteries, which branch and narrow into the arterioles, and then branch further still into the capillaries. After the tissue has been perfused, capillaries join and widen to become venules and then widen more to become veins, which return blood to the heart.
	Capillaries do not function on their own. The 'capillary bed' is an interweaving network of capillaries supplying an organ.
Kidney development	Kidney development, describes the embryologic origins of the kidney, a major organ in the urinary system. It is often considered in the broader context of the development of the urinary and reproductive organs.
	The development of the kidney proceeds through a series of successive phases, each marked by the development of a more advanced kidney: the pronephros, mesonephros, and metanephros.
Urinary system	The urinary system is the organ system that produces, stores, and eliminates urine. In humans it includes two kidneys, two ureters, the bladder, the urethra, and two sphincter muscles.
Renal agenesis	Renal agenesis is a medical condition in which one (unilateral) or both (bilateral) fetal kidneys fail to develop.
	It can be associated with RET or UPK3A.Bilateral

Chapter 9. Embryo development and fetal growth

CHAPTER HIGHLIGHTS & NOTES: KEY TERMS, PEOPLE, PLACES, CONCEPTS

	Bilateral renal agenesis is the uncommon and serious failure of both a fetus' kidneys to develop during gestation, and is one causative agent of Potter sequence. This absence of kidneys causes oligohydramnios, a deficiency of amniotic fluid in a pregnant woman, which can place extra pressure on the developing baby and cause further malformations.
Amniotic fluid	Amniotic fluid is the nourishing and protecting liquid contained by the amniotic sac of a pregnant woman.
	From the very beginning of the formation of the extracoelomal cavity, amniotic fluid [AF] can be detected. This firstly water-like fluid originates from the maternal plasma, and passes through the fetal membranes by osmotic and hydrostatic forces.
Small for gestational age	Small for gestational age babies are those who are smaller in size than normal for the baby's sex and gestational age, most commonly defined as a weight below the 10th percentile for the gestational age.
	Not all fetuses that are SGA are pathologically growth restricted and, in fact, may be constitutionally small. If small for gestational age babies have been the subject of intrauterine growth restriction (IUGR), formerly known as intrauterine growth retardation, the term SGA associated with IUGR is used.
Gestational age	Gestational age is the age of an embryo or fetus . In humans, a common method of calculating gestational age starts counting either from the first day of the woman's last menstrual period (LMP) or from 14 days before conception (fertilization). Counting from the first day of the LMP involves the assumption that conception occurred 14 days later.
Growth hormone	Growth hormone is a peptide hormone that stimulates growth, cell reproduction and regeneration in humans and other animals. Growth hormone is a 191-amino acid, single-chain polypeptide that is synthesized, stored, and secreted by the somatotroph cells within the lateral wings of the anterior pituitary gland. Somatotropin (STH) refers to the growth hormone 1 produced naturally in animals, whereas the term somatropin refers to growth hormone produced by recombinant DNA technology, and is abbreviated 'HGH' in humans.
Growth factor	A growth factor is a naturally occurring substance capable of stimulating cellular growth, proliferation and cellular differentiation. Usually it is a protein or a steroid hormone. Growth factors are important for regulating a variety of cellular processes.
Glucocorticoid	Glucocorticoids are a class of steroid hormones that bind to the glucocorticoid receptor (GR), which is present in almost every vertebrate animal cell.

Chapter 9. Embryo development and fetal growth

CHAPTER HIGHLIGHTS & NOTES: KEY TERMS, PEOPLE, PLACES, CONCEPTS

	Glucocorticoids are part of the feedback mechanism in the immune system that turns immune activity (inflammation) down. They are therefore used in medicine to treat diseases that are caused by an overactive immune system, such as allergies, asthma, autoimmune diseases and sepsis. Glucocorticoids have many diverse (pleiotropic) effects, including potentially harmful side effects, and as a result are rarely sold over-the-counter. They also interfere with some of the abnormal mechanisms in cancer cells, so they are used in high doses to treat cancer.
Maternal nutrition	Maternal nutrition is the dietary intake and habits of expectant mothers with dual emphasis on the health of the mother and the physical and mental development of infants. Nearly 24 per cent of babies are estimated to be born with lower than optimal weights at birth. Pregnancy and child birth transform every aspect of expecting mother's lives. They should be very careful with their choices regarding what they eat.
Infection	An infection is the colonization of a host organism by parasite species. Infecting parasites seek to use the host's resources to reproduce, often resulting in disease. Colloquially, infections are usually considered to be caused by microscopic organisms or microparasites like viruses, prions, bacteria, and viroids, though larger organisms like macroparasites and fungi can also infect.
Obesity	Obesity is a medical condition in which excess body fat has accumulated to the extent that it may have an adverse effect on health, leading to reduced life expectancy and/or increased health problems. Body mass index (BMI), a measurement which compares weight and height, defines people as overweight (pre-obese) if their BMI is between 25 and 30 kg/m^2, and obese when it is greater than 30 kg/m^2. Obesity increases the likelihood of various diseases, particularly heart disease, type 2 diabetes, obstructive sleep apnea, certain types of cancer, and osteoarthritis.
Pre-eclampsia	Pre-eclampsia is a medical condition in which hypertension arises in pregnancy (pregnancy-induced hypertension) in association with significant amounts of protein in the urine. Pre-eclampsia refers to a set of symptoms rather than any causative factor, and there are many different causes for the condition. It appears likely that there are substances from the placenta that can cause endothelial dysfunction in the maternal blood vessels of susceptible women.
Immune system	The immune system is a system of biological structures and processes within an organism that protects against disease. To function properly, an immune system must detect a wide variety of agents, from viruses to parasitic worms, and distinguish them from the organism's own healthy tissue.

Chapter 9. Embryo development and fetal growth

CHAPTER QUIZ: KEY TERMS, PEOPLE, PLACES, CONCEPTS

1. _____ is the process of blood vessel formation occurring by a de novo production of endothelial cells.

   Though similar to angiogenesis, the two are different in one aspect: The term angiogenesis denotes the formation of new blood vessels from pre-existing ones, whereas _____ is the term used for the formation of new blood vessels when there are no pre-existing ones. For example, if a monolayer of endothelial cells begins sprouting to form capillaries, angiogenesis is occurring.

   a. Vitelline arteries
   b. Vitelline veins
   c. Vasculogenesis
   d. Primary interventricular foramen

2. The _____ is the membrane that covers the embryonic cloaca when still in the development of the urinary and reproductive organs.

   It is formed by ectoderm and endoderm coming into contact with each other. After separation of the cloaca into the urogenital and anal parts, the _____, in turn, is separated into a urogenital membrane and an anal membrane.

   a. Foregut
   b. Cloacal membrane
   c. Hindgut
   d. Midgut

3. _____ is the process of programmed cell death (PCD) that may occur in multicellular organisms. Biochemical events lead to characteristic cell changes (morphology) and death. These changes include blebbing, cell shrinkage, nuclear fragmentation, chromatin condensation, and chromosomal DNA fragmentation.

   a. Idiotopes
   b. IRGs
   c. Apoptosis
   d. Febrile neutrophilic dermatosis

4. The _____s (ductus ejaculatorii) are paired structures in male anatomy. Each _____ is formed by the union of the vas deferens with the duct of the seminal vesicle. They pass through the prostate, and open into the urethra at the Colliculus seminalis.

   a. Excretory duct of seminal gland
   b. External spermatic fascia
   c. External urethral orifice
   d. Ejaculatory duct

5. . The _____ is the biological system of an organism that introduces respiratory gases to the interior and performs gas exchange.

Chapter 9. Embryo development and fetal growth

CHAPTER QUIZ: KEY TERMS, PEOPLE, PLACES, CONCEPTS

In humans and other mammals, the anatomical features of the _____ include airways, lungs, and the respiratory muscles. Molecules of oxygen and carbon dioxide are passively exchanged, by diffusion, between the gaseous external environment and the blood.

a. Respiratory system
b. Posterior cardinal vein
c. Primary interatrial foramen
d. Primary interventricular foramen

ANSWER KEY
Chapter 9. Embryo development and fetal growth

1. c
2. b
3. c
4. d
5. a

You can take the complete Chapter Practice Test

for Chapter 9. Embryo development and fetal growth
on all key terms, persons, places, and concepts.

Online 99 Cents

http://www.epub5042.32.20612.9.cram101.com/

Use www.Cram101.com for all your study needs

including Cram101's online interactive problem solving labs in

chemistry, statistics, mathematics, and more.

Chapter 10. Overview of immunology

CHAPTER OUTLINE: KEY TERMS, PEOPLE, PLACES, CONCEPTS

	Immunology
	Immune system
	Respiratory system
	Complement system
	Interferon
	Toll-like receptor
	Granulocyte
	Lymphocyte
	Macrophage
	Monocyte
	Natural killer cell
	Neutrophil granulocyte
	Opsonin
	Phagocyte
	Stem cell
	Central tolerance
	Clonal deletion
	T cell
	Clonal selection

Chapter 10. Overview of immunology
CHAPTER OUTLINE: KEY TERMS, PEOPLE, PLACES, CONCEPTS

_____ Passive immunity

_____ Antigen-presenting cell

_____ Cytotoxic T cell

_____ T helper cell

_____ Regulatory T cell

_____ Major histocompatibility complex

_____ Fetus

_____ Hypertension

_____ Uterus

_____ Abortion

_____ Autoimmune disease

_____ Rectum

_____ Interleukin

_____ Rheumatoid arthritis

_____ Tryptophan

_____ Infection

_____ Microchimerism

_____ Active immunization

_____ Immunization

Chapter 10. Overview of immunology

CHAPTER OUTLINE: KEY TERMS, PEOPLE, PLACES, CONCEPTS

	Large intestine
	Endometriosis
	Pre-eclampsia

CHAPTER HIGHLIGHTS & NOTES: KEY TERMS, PEOPLE, PLACES, CONCEPTS	
Immunology	Immunology is a branch of biomedical science that covers the study of all aspects of the immune system in all organisms. It deals with the physiological functioning of the immune system in states of both health and diseases; malfunctions of the immune system in immunological disorders (autoimmune diseases, hypersensitivities, immune deficiency, transplant rejection); the physical, chemical and physiological characteristics of the components of the immune system in vitro, in situ, and in vivo. Immunology has applications in several disciplines of science, and as such is further divided.
Immune system	The immune system is a system of biological structures and processes within an organism that protects against disease. To function properly, an immune system must detect a wide variety of agents, from viruses to parasitic worms, and distinguish them from the organism's own healthy tissue. Pathogens can rapidly evolve and adapt to avoid detection and neutralization by the immune system.
Respiratory system	The respiratory system is the biological system of an organism that introduces respiratory gases to the interior and performs gas exchange. In humans and other mammals, the anatomical features of the respiratory system include airways, lungs, and the respiratory muscles. Molecules of oxygen and carbon dioxide are passively exchanged, by diffusion, between the gaseous external environment and the blood.
Complement system	The complement system helps or 'complements' the ability of antibodies and phagocytic cells to clear pathogens from an organism. It is part of the immune system called the innate immune system that is not adaptable and does not change over the course of an individual's lifetime. However, it can be recruited and brought into action by the adaptive immune system.

Chapter 10. Overview of immunology

CHAPTER HIGHLIGHTS & NOTES: KEY TERMS, PEOPLE, PLACES, CONCEPTS

Interferon	Interferon alpha/beta domain
	Interferons (IFNs) are proteins made and released by host cells in response to the presence of pathogens such as viruses, bacteria, parasites or tumor cells. They allow for communication between cells to trigger the protective defenses of the immune system that eradicate pathogens or tumors.
	IFNs belong to the large class of glycoproteins known as cytokines.
Toll-like receptor	Toll-like receptors (TLRs) are a class of proteins that play a key role in the innate immune system. They are single, membrane-spanning, non-catalytic receptors that recognize structurally conserved molecules derived from microbes. Once these microbes have breached physical barriers such as the skin or intestinal tract mucosa, they are recognized by TLRs, which activate immune cell responses.
Granulocyte	Granulocytes are a category of white blood cells characterized by the presence of granules in their cytoplasm. They are also called polymorphonuclear leukocytes (PMN or PML) because of the varying shapes of the nucleus, which is usually lobed into three segments. In common parlance, the term polymorphonuclear leukocyte often refers specifically to neutrophil granulocytes, the most abundant of the granulocytes.
Lymphocyte	A lymphocyte is a type of white blood cell in the vertebrate immune system.
	Under the microscope, lymphocytes can be divided into large granular lymphocytes and small lymphocytes. Large granular lymphocytes include natural killer cells (NK cells). Small lymphocytes consist of T cells and B cells.
Macrophage	Macrophages are cells produced by the differentiation of monocytes in tissues. Human macrophages are about 21 micrometres (0.00083 in) in diameter. Monocytes and macrophages are phagocytes.
Monocyte	Monocyte is a type of white blood cell and is part of the human body's immune system. Monocytes play multiple roles in immune function. Such roles include: (1) replenish resident macrophages and dendritic cells under normal states, and (2) in response to inflammation signals, monocytes can move quickly (approx. 8-12 hours) to sites of infection in the tissues and divide/differentiate into macrophages and dendritic cells to elicit an immune response. Half of them are stored in the spleen. Monocytes are usually identified in stained smears by their large kidney shaped or notched nucleus.
Natural killer cell	Natural killer cells are a type of cytotoxic lymphocyte that constitute a major component of the innate immune system.

Chapter 10. Overview of immunology

CHAPTER HIGHLIGHTS & NOTES: KEY TERMS, PEOPLE, PLACES, CONCEPTS

	NK cells play a major role in the rejection of tumors and cells infected by viruses. They kill cells by releasing small cytoplasmic granules of proteins called perforin and granzyme that cause the target cell to die by apoptosis (programmed cell death).
Neutrophil granulocyte	Neutrophil granulocytes are generally referred to as either neutrophils or polymorphonuclear neutrophils (or PMNs), and are subdivided into segmented neutrophils (or segs) and banded neutrophils (or bands). Neutrophils are the most abundant type of white blood cells in mammals and form an essential part of the innate immune system. They form part of the polymorphonuclear cell family (PMNs) together with basophils and eosinophils. Whereas basophilic white blood cells stain dark blue and eosinophilic white blood cells stain bright red, neutrophils stain a neutral pink. Normally neutrophils contain a nucleus divided into 2-5 lobes.
Opsonin	An opsonin is any molecule that targets an antigen for an immune response. However, the term is usually used in reference to molecules that act as a binding enhancer for the process of phagocytosis, especially antibodies, which coat the negatively-charged molecules on the membrane. Molecules that activate the complement system are also considered opsonins.
Phagocyte	Phagocytes are the white blood cells that protect the body by ingesting (phagocytosing) harmful foreign particles, bacteria, and dead or dying cells. Their name comes from the Greek phagein, 'to eat' or 'devour', and '-cyte', the suffix in biology denoting 'cell', from the Greek kutos, 'hollow vessel'. They are essential for fighting infections and for subsequent immunity.
Stem cell	Stem cells are biological cells found in all multicellular organisms, that can divide (through mitosis) and differentiate into diverse specialized cell types and can self-renew to produce more stem cells. In mammals, there are two broad types of stem cells: embryonic stem cells, which are isolated from the inner cell mass of blastocysts, and adult stem cells, which are found in various tissues. In adult organisms, stem cells and progenitor cells act as a repair system for the body, replenishing adult tissues.
Central tolerance	Central tolerance is the mechanism by which newly developing T cells and B cells are rendered non-reactive to self. The concept of central tolerance was proposed in 1959 by Joshua Lederberg, as part of his general theory of immunity and tolerance, and is often mistakenly attributed to MacFarlane Burnet. Lederberg hypothesized that it is the age of the lymphocyte that defines whether an antigen that is encountered will induce tolerance, with immature lymphocytes being tolerance sensitive.
Clonal deletion	Clonal deletion is a process by which B cells and T cells are deactivated after they have expressed receptors for self-antigens and before they develop into fully immunocompetent lymphocytes.

Chapter 10. Overview of immunology

CHAPTER HIGHLIGHTS & NOTES: KEY TERMS, PEOPLE, PLACES, CONCEPTS

T cell	T cells belong to a group of white blood cells known as lymphocytes, and play a central role in cell-mediated immunity. They can be distinguished from other lymphocyte types, such as B cells and natural killer cells (NK cells) by the presence of a special receptor on their cell surface called T cell receptors (TCR). The abbreviation T, in T cell, stands for thymus, since this is the principal organ responsible for the T cell's maturation. Several different subsets of T cells have been discovered, each with a distinct function.
Clonal selection	The clonal selection hypothesis has become a widely accepted model for how the immune system responds to infection and how certain types of B and T lymphocytes are selected for destruction of specific antigens invading the body. •Each lymphocyte bears a single type of receptor with a unique specificity (by V(D)J recombination).•Receptor occupation is required for cell activation.•The differentiated effector cells derived from an activated lymphocyte will bear receptors of identical specificity as the parental cell.•Those lymphocytes bearing receptors for self molecules will be deleted at an early stage.Early work
In 1954, Danish immunologist Niels Jerne put forward a hypothesis which stated that there is already a vast array of lymphocytes in the body prior to any infection. The entrance of an antigen into the body results in the selection of only one type of lymphocyte to match it and produce a corresponding antibody to destroy the antigen.
Passive immunity	Passive immunity is the transfer of active humoral immunity in the form of readymade antibodies, from one individual to another. Passive immunity can occur naturally, when maternal antibodies are transferred to the fetus through the placenta, and can also be induced artificially, when high levels of human antibodies specific for a pathogen or toxin are transferred to non-immune individuals. Passive immunization is used when there is a high risk of infection and insufficient time for the body to develop its own immune response, or to reduce the symptoms of ongoing or immunosuppressive diseases.
Antigen-presenting cell	An antigen-presenting cell is a cell that displays foreign antigen complexes with major histocompatibility complex (MHC) on their surfaces. T-cells may recognize these complexes using their T-cell receptors (TCRs). These cells process antigens and present them to T-cells.
Cytotoxic T cell	A cytotoxic T cell belongs to a sub-group of T lymphocytes (a type of white blood cell) that are capable of inducing the death of infected somatic or tumor cells; they kill cells that are infected with viruses (or other pathogens), or are otherwise damaged or dysfunctional. Most cytotoxic T cells express T-cell receptors (TCRs) that can recognize a specific antigenic peptide bound to Class I MHC molecules, present on all nucleated cells, and a glycoprotein called CD8, which is attracted to non-variable portions of the Class I MHC molecule. The affinity between CD8 and the MHC molecule keeps the T_C cell and the target cell bound closely together during antigen-specific activation.

Chapter 10. Overview of immunology

CHAPTER HIGHLIGHTS & NOTES: KEY TERMS, PEOPLE, PLACES, CONCEPTS

T helper cell	T helper cells are a sub-group of lymphocytes, a type of white blood cell, that play an important role in establishing and maximizing the capabilities of the immune system. These cells are unusual in that they have no cytotoxic or phagocytic activity; they cannot kill infected host cells (also known as somatic cells) or pathogens, and without other immune cells they would usually be considered useless against an infection. T_h cells are involved in activating and directing other immune cells, and are particularly important in the immune system. They are essential in determining B cell antibody class switching, in the activation and growth of cytotoxic T cells, and in maximizing bactericidal activity of phagocytes such as macrophages. It is this diversity in function and their role in influencing other cells that gives T helper cells their name.
Regulatory T cell	Regulatory T cells are a specialized subpopulation of T cells that act to suppress activation of the immune system and thereby maintain immune system homeostasis and tolerance to self-antigens. The existence of a dedicated population of suppressive T cells was the subject of significant controversy among immunologists for many years. However, recent advances in the molecular characterization of this cell population have firmly established their existence and their critical role in the vertebrate immune system. Interest in regulatory T cells has been heightened by evidence from experimental mouse models demonstrating that the immunosuppressive potential of these cells can be harnessed therapeutically to treat autoimmune diseases and facilitate transplantation tolerance or specifically eliminated to potentiate cancer immunotherapy.
Major histocompatibility complex	Major histocompatibility complex is a cell surface molecule encoded by a large gene family in all vertebrates. MHC molecules mediate interactions of leukocytes, also called white blood cells (WBCs), which are immune cells, with other leukocytes or body cells. MHC determines compatibility of donors for organ transplant as well as one's susceptibility to an autoimmune disease via crossreacting immunization.
Fetus	A fetus (sometimes spelled foetus or fœtus) is a stage in the development of viviparous organisms. This stage lies between the embryonic stage and birth. The fetuses of most mammals are situated similarly to the homo sapiens fetus within their mothers.
Hypertension	Hypertension or high blood pressure, sometimes called arterial hypertension, is a chronic medical condition in which the blood pressure in the arteries is elevated. This requires the heart to work harder than normal to circulate blood through the blood vessels. Blood pressure involves two measurements, systolic and diastolic, which depend on whether the heart muscle is contracting (systole) or relaxed between beats (diastole).
Uterus	The uterus is a major female hormone-responsive reproductive sex organ of most mammals including humans. One end, the cervix, opens into the vagina, while the other is connected to one or both fallopian tubes, depending on the species.

Chapter 10. Overview of immunology

CHAPTER HIGHLIGHTS & NOTES: KEY TERMS, PEOPLE, PLACES, CONCEPTS

Abortion	Abortion is defined as the termination of pregnancy by the removal or expulsion from the uterus of a fetus or embryo prior to viability. An abortion can occur spontaneously, in which case it is usually called a miscarriage, or it can be purposely induced. The term abortion most commonly refers to the induced abortion of a human pregnancy.
Autoimmune disease	Autoimmune diseases arise from an inappropriate immune response of the body against substances and tissues normally present in the body. In other words, the immune system mistakes some part of the body as a pathogen and attacks its own cells. This may be restricted to certain organs (e.g. in autoimmune thyroiditis) or involve a particular tissue in different places (e.g. Goodpasture's disease which may affect the basement membrane in both the lung and the kidney).
Rectum	The rectum is the final straight portion of the large intestine in some mammals, and the gut in others, terminating in the anus. The human rectum is about 12 centimetres (4.7 in) long. Its caliber is similar to that of the sigmoid colon at its commencement, but it is dilated near its termination, forming the rectal ampulla.
Interleukin	Interleukins are a group of cytokines (secreted proteinssignaling molecules) that were first seen to be expressed by white blood cells (leukocytes). The term interleukin derives from (inter-) 'as a means of communication', and (-leukin) 'deriving from the fact that many of these proteins are produced by leukocytes and act on leukocytes'. The name is something of a relic, though (the term was coined by Dr. Vern Paetkau, University of Victoria); it has since been found that interleukins are produced by a wide variety of body cells.
Rheumatoid arthritis	Rheumatoid arthritis is a chronic, systemic inflammatory disorder that may affect many tissues and organs, but principally attacks flexible (synovial) joints. The process involves an inflammatory response of the capsule around the joints (synovium) secondary to swelling (hyperplasia) of synovial cells, excess synovial fluid, and the development of fibrous tissue (pannus) in the synovium. The pathology of the disease process often leads to the destruction of articular cartilage and ankylosis (fusion) of the joints.
Tryptophan	Tryptophan is one of the 20 standard amino acids, as well as an essential amino acid in the human diet. It is encoded in the standard genetic code as the codon UGG. The slight mispronunciation 'tWiptophan' can be used as a mnemonic for its single letter IUPAC code W. Only the L-stereoisomer of tryptophan is used in structural or enzyme proteins, but the D-stereoisomer is occasionally found in naturally produced peptides (for example, the marine venom peptide contryphan). The distinguishing structural characteristic of tryptophan is that it contains an indole functional group.
Infection	An infection is the colonization of a host organism by parasite species. Infecting parasites seek to use the host's resources to reproduce, often resulting in disease.

Chapter 10. Overview of immunology

CHAPTER HIGHLIGHTS & NOTES: KEY TERMS, PEOPLE, PLACES, CONCEPTS

Microchimerism	Microchimerism is the presence of a small number of cells that originate from another individual and are therefore genetically distinct from the cells of the host individual. This phenomenon may be related to certain types of autoimmune diseases; however, the mechanisms responsible for this relationship are unclear. Human
	In humans (and perhaps in all Placentals) the most common form is fetomaternal microchimerism whereby cells from a fetus pass through the placenta and establish cell lineages within the mother.
Active immunization	Active immunization is the induction of immunity after exposure to an antigen. Antibodies are created by the recipient and may be stored permanently.
	Active immunization can occur naturally when a microbe or other antigen is received by a person who has not yet come into contact with the microbe and has no pre-made antibodies for defense.
Immunization	Immunization, is the process by which an individual's immune system becomes fortified against an agent (known as the immunogen).
	When this system is exposed to molecules that are foreign to the body, called non-self, it will orchestrate an immune response, and it will also develop the ability to quickly respond to a subsequent encounter because of immunological memory. This is a function of the adaptive immune system.
Large intestine	The large intestine is the third-to-last part of the digestive system in vertebrate animals.
Endometriosis	Endometriosis is a gynecological medical condition in which cells from the lining of the uterus (endometrium) appear and flourish outside the uterine cavity, most commonly on the ovaries. The uterine cavity is lined by endometrial cells, which are under the influence of female hormones. These endometrial-like cells in areas outside the uterus (endometriosis) are influenced by hormonal changes and respond in a way that is similar to the cells found inside the uterus.
Pre-eclampsia	Pre-eclampsia is a medical condition in which hypertension arises in pregnancy (pregnancy-induced hypertension) in association with significant amounts of protein in the urine.
	Pre-eclampsia refers to a set of symptoms rather than any causative factor, and there are many different causes for the condition. It appears likely that there are substances from the placenta that can cause endothelial dysfunction in the maternal blood vessels of susceptible women.

Chapter 10. Overview of immunology

CHAPTER QUIZ: KEY TERMS, PEOPLE, PLACES, CONCEPTS

1. _____ is a branch of biomedical science that covers the study of all aspects of the immune system in all organisms. It deals with the physiological functioning of the immune system in states of both health and diseases; malfunctions of the immune system in immunological disorders (autoimmune diseases, hypersensitivities, immune deficiency, transplant rejection); the physical, chemical and physiological characteristics of the components of the immune system in vitro, in situ, and in vivo. _____ has applications in several disciplines of science, and as such is further divided.

    a. Immunology
    b. Addiction module
    c. Adoptive immunity
    d. Allele frequency net database

2. _____ is a type of white blood cell and is part of the human body's immune system. _____s play multiple roles in immune function. Such roles include: (1) replenish resident macrophages and dendritic cells under normal states, and (2) in response to inflammation signals, _____s can move quickly (approx. 8-12 hours) to sites of infection in the tissues and divide/differentiate into macrophages and dendritic cells to elicit an immune response. Half of them are stored in the spleen. _____s are usually identified in stained smears by their large kidney shaped or notched nucleus.

    a. pre-B cell
    b. MHC class I
    c. Monocyte
    d. MHC restriction

3. The _____ is the biological system of an organism that introduces respiratory gases to the interior and performs gas exchange. In humans and other mammals, the anatomical features of the _____ include airways, lungs, and the respiratory muscles. Molecules of oxygen and carbon dioxide are passively exchanged, by diffusion, between the gaseous external environment and the blood.

    a. magnetic resonance imaging
    b. Immunity
    c. Immunization
    d. Respiratory system

4. . A _____ belongs to a sub-group of T lymphocytes (a type of white blood cell) that are capable of inducing the death of infected somatic or tumor cells; they kill cells that are infected with viruses (or other pathogens), or are otherwise damaged or dysfunctional. Most _____s express T-cell receptors (TCRs) that can recognize a specific antigenic peptide bound to Class I MHC molecules, present on all nucleated cells, and a glycoprotein called CD8, which is attracted to non-variable portions of the Class I MHC molecule. The affinity between CD8 and the MHC molecule keeps the T_C cell and the target cell bound closely together during antigen-specific activation.

    a. Natural killer cell
    b. Cohort effect
    c. Collider

Chapter 10. Overview of immunology

CHAPTER QUIZ: KEY TERMS, PEOPLE, PLACES, CONCEPTS

5. _____ is defined as the termination of pregnancy by the removal or expulsion from the uterus of a fetus or embryo prior to viability. An _____ can occur spontaneously, in which case it is usually called a miscarriage, or it can be purposely induced. The term _____ most commonly refers to the induced _____ of a human pregnancy.

a. Adultery
b. Abortion
c. American Fertility Association
d. Assortative mating

ANSWER KEY
Chapter 10. Overview of immunology

1. a
2. c
3. d
4. d
5. b

You can take the complete Chapter Practice Test

for Chapter 10. Overview of immunology
on all key terms, persons, places, and concepts.

Online 99 Cents

http://www.epub5042.32.20612.10.cram101.com/

Use www.Cram101.com for all your study needs

including Cram101's online interactive problem solving labs in

chemistry, statistics, mathematics, and more.

Chapter 11. Physiological adaptation to pregnancy

CHAPTER OUTLINE: KEY TERMS, PEOPLE, PLACES, CONCEPTS

- Endocrine system
- Fetus
- Immune system
- Steroid hormone
- Ectopic pregnancy
- Gonadotropin
- Adrenocorticotropic hormone
- Cortisol
- Human placental lactogen
- Placental lactogen
- Hypothyroidism
- Thyroid hormone
- Thyroxine
- Triiodothyronine
- Braxton Hicks contractions
- Reproductive system
- Myocardial infarction
- Kidney
- Renin-angiotensin system

Chapter 11. Physiological adaptation to pregnancy
CHAPTER OUTLINE: KEY TERMS, PEOPLE, PLACES, CONCEPTS

	Volume
	Heart development
	Blood pressure
	Nitric oxide
	Varicose veins
	Monitoring
	Blood flow
	Blood cell
	Platelet
	Diaphragm
	Respiratory system
	Diffusion capacity
	Partial pressure
	Prostaglandin
	Respiratory alkalosis
	Respiratory rate
	Oxygen
	Ureter
	Urinary incontinence

Chapter 11. Physiological adaptation to pregnancy

CHAPTER OUTLINE: KEY TERMS, PEOPLE, PLACES, CONCEPTS

	Urinary system
	Urinary tract infection
	Progesterone
	Food craving
	Hyperemesis gravidarum
	Nausea
	Vomiting
	Large intestine
	Small intestine
	Calcium metabolism
	Vitamin D
	Metabolism
	Skeleton
	Larynx
	Diabetes mellitus
	Gestational diabetes

Chapter 11. Physiological adaptation to pregnancy

CHAPTER HIGHLIGHTS & NOTES: KEY TERMS, PEOPLE, PLACES, CONCEPTS

Endocrine system	The endocrine system is the system of glands, each of which secretes a type of hormone directly into the bloodstream to regulate the body. The endocrine system is in contrast to the exocrine system, which secretes its chemicals using ducts. It derives from the Greek words 'endo' meaning inside, within, and 'crinis' for secrete.
Fetus	A fetus (sometimes spelled foetus or fœtus) is a stage in the development of viviparous organisms. This stage lies between the embryonic stage and birth. The fetuses of most mammals are situated similarly to the homo sapiens fetus within their mothers.
Immune system	The immune system is a system of biological structures and processes within an organism that protects against disease. To function properly, an immune system must detect a wide variety of agents, from viruses to parasitic worms, and distinguish them from the organism's own healthy tissue. Pathogens can rapidly evolve and adapt to avoid detection and neutralization by the immune system.
Steroid hormone	A steroid hormone is a steroid that acts as a hormone. Steroid hormones can be grouped into five groups by the receptors to which they bind: glucocorticoids, mineralocorticoids, androgens, estrogens, and progestogens. Vitamin D derivatives are a sixth closely related hormone system with homologous receptors, though technically sterols rather than steroids.
Ectopic pregnancy	An ectopic pregnancy, is a complication of pregnancy in which the embryo implants outside the uterine cavity. With rare exceptions, ectopic pregnancies are not viable. Furthermore, they are dangerous for the parent, since internal haemorrhage is a life threatening complication.
Gonadotropin	Glycoprotein hormone Gonadotropins are protein hormones secreted by gonadotrope cells of the pituitary gland of vertebrates. This is a family of proteins, which include the mammalian hormones follitropin (FSH), lutropin (LH), placental chorionic gonadotropins hCG and eCG and chorionic gonadotropin as well as at least two forms of fish gonadotropins. These hormones are central to the complex endocrine system that regulates normal growth, sexual development, and reproductive function.
Adrenocorticotropic hormone	Adrenocorticotropic hormone also known as corticotropin, is a polypeptide tropic hormone produced and secreted by the anterior pituitary gland. It is an important component of the hypothalamic-pituitary-adrenal axis and is often produced in response to biological stress (along with its precursor corticotropin-releasing hormone from the hypothalamus).

Chapter 11. Physiological adaptation to pregnancy

CHAPTER HIGHLIGHTS & NOTES: KEY TERMS, PEOPLE, PLACES, CONCEPTS

Cortisol	Cortisol is a steroid hormone, or glucocorticoid, produced by the adrenal gland. It is released in response to stress and a low level of blood glucocorticoids. Its primary functions are to increase blood sugar through gluconeogenesis; suppress the immune system; and aid in fat, protein and carbohydrate metabolism.
Human placental lactogen	Human placental lactogen also called human chorionic somatomammotropin, is a polypeptide placental hormone. Its structure and function is similar to that of human growth hormone. It modifies the metabolic state of the mother during pregnancy to facilitate the energy supply of the fetus.
Placental lactogen	Placental lactogen is a polypeptide placental hormone. Its structure and function is similar to that of growth hormone. It modifies the metabolic state of the mother during pregnancy to facilitate the energy supply of the fetus.
Hypothyroidism	Hypothyroidism is a condition in which the thyroid gland does not make enough thyroid hormone. Iodine deficiency is often cited as the most common cause of hypothyroidism worldwide but it can be caused by many other factors. It can result from a lack of a thyroid gland or from iodine-131 treatment, and can also be associated with increased stress.
Thyroid hormone	The thyroid hormones, thyroxine (T_4) and triiodothyronine (T_3), are tyrosine-based hormones produced by the thyroid gland primarily responsible for regulation of metabolism. An important component in the synthesis of thyroid hormones is iodine. The major form of thyroid hormone in the blood is thyroxine (T_4), which has a longer half life than T_3.
Thyroxine	Thyroxine, or 3,5,3',5'-tetraiodothyronine, a form of thyroid hormones is the major hormone secreted by the follicular cells of the thyroid gland. Synthesis and regulation Thyroxine is synthesized via the iodination and covalent bonding of the phenyl portions of tyrosine residues found in an initial peptide, thyroglobulin, which is secreted into thyroid granules. These iodinated diphenyl compounds are cleaved from their peptide backbone upon being stimulated by thyroid-stimulating hormone.
Triiodothyronine	Triiodothyronine, $C_{15}H_{12}I_3NO_4$, also known as T_3, is a thyroid hormone. It affects almost every physiological process in the body, including growth and development, metabolism, body temperature, and heart rate.

Chapter 11. Physiological adaptation to pregnancy

CHAPTER HIGHLIGHTS & NOTES: KEY TERMS, PEOPLE, PLACES, CONCEPTS

Braxton Hicks contractions	Braxton Hicks contractions, are sporadic uterine contractions that sometimes start around 6 weeks. However, they are not usually felt until the second trimester or third trimester of pregnancy. They should be infrequent, irregular, and involve only mild cramping.
Reproductive system	The reproductive system is a system of organs within an organism which work together for the purpose of reproduction. Many non-living substances such as fluids, hormones, and pheromones are also important accessories to the reproductive system. Unlike most organ systems, the sexes of differentiated species often have significant differences.
Myocardial infarction	Myocardial infarction or acute myocardial infarction commonly known as a heart attack, results from the interruption of blood supply to a part of the heart, causing heart cells to die. This is most commonly due to occlusion (blockage) of a coronary artery following the rupture of a vulnerable atherosclerotic plaque, which is an unstable collection of lipids (cholesterol and fatty acids) and white blood cells (especially macrophages) in the wall of an artery. The resulting ischemia (restriction in blood supply) and ensuing oxygen shortage, if left untreated for a sufficient period of time, can cause damage or death (infarction) of heart muscle tissue (myocardium).
Kidney	The kidneys are organs with several functions. They are seen in many types of animals, including vertebrates and some invertebrates. They are an essential part of the urinary system and also serve homeostatic functions such as the regulation of electrolytes, maintenance of acid-base balance, and regulation of blood pressure. They serve the body as a natural filter of the blood, and remove wastes which are diverted to the urinary bladder. In producing urine, the kidneys excrete wastes such as urea and ammonium; the kidneys also are responsible for the reabsorption of water, glucose, and amino acids. The kidneys also produce hormones including calcitriol, renin, and erythropoietin.
Renin-angiotensin system	The renin-angiotensin system or the renin-angiotensin-aldosterone system (RAAS) is a hormone system that regulates blood pressure and water (fluid) balance. When blood volume is low, juxtaglomerular cells in the kidneys secrete renin. Renin stimulates the production of angiotensin I, which is then converted to angiotensin II. Angiotensin II causes blood vessels to constrict, resulting in increased blood pressure.
Volume	Volume is how much three-dimensional space a substance (solid, liquid, gas, or plasma) or shape occupies or contains, often quantified numerically using the SI derived unit, the cubic metre. The volume of a container is generally understood to be the capacity of the container, i. e. the amount of fluid (gas or liquid) that the container could hold, rather than the amount of space the container itself displaces.

Chapter 11. Physiological adaptation to pregnancy

CHAPTER HIGHLIGHTS & NOTES: KEY TERMS, PEOPLE, PLACES, CONCEPTS

Heart development	The heart is the first functional organ in a vertebrate embryo. There are 5 stages to heart development. The lateral plate mesoderm delaminates to form two layers: the dorsal somatic (parietal) mesoderm and the ventral splanchnic (visceral) mesoderm.
Blood pressure	Blood pressure is the pressure exerted by circulating blood upon the walls of blood vessels, and is one of the principal vital signs. When used without further specification, 'blood pressure' usually refers to the arterial pressure of the systemic circulation. During each heartbeat, blood pressure varies between a maximum (systolic) and a minimum (diastolic) pressure.
Nitric oxide	Nitric oxide is a chemical compound with chemical formula Nitric oxide. This diatomic gas is an important cell signaling molecule in mammals, including humans. It is also an extremely important intermediate in the chemical industry. Nitric oxide is an air pollutant produced by combustion of substances in air, like in automobile engines and fossil fuel power plants.
Varicose veins	Varicose veins are veins that have become enlarged and tortuous. The term commonly refers to the veins on the leg, although varicose veins can occur elsewhere. Veins have leaflet valves to prevent blood from flowing backwards (retrograde flow or reflux).
Monitoring	In medicine, monitoring is the evaluation of a disease or condition over time. It can be performed by continuously measuring certain parameters (for example, by continuously measuring vital signs by a bedside monitor), and/or by repeatedly performing medical tests (such as blood glucose monitoring in people with diabetes mellitus). Transmitting data from a monitor to a distant monitoring station is known as telemetry or biotelemetry.
Blood flow	Blood flow is the continuous running of blood in the cardiovascular system. The human body is made up of several processes all carrying out various functions. We have the gastrointestinal system which aids the digestion and the absorption of food.
Blood cell	A blood cell, is a cell produced by haematopoiesis and normally found in blood. In mammals, these fall into three general categories:•Red blood cells -- Erythrocytes•White blood cells -- Leukocytes•Platelets -- Thrombocytes. Together, these three kinds of blood cells add up to a total 45% of the blood tissue by volume, with the remaining 55% of the volume composed of plasma, the liquid component of blood.

Chapter 11. Physiological adaptation to pregnancy

CHAPTER HIGHLIGHTS & NOTES: KEY TERMS, PEOPLE, PLACES, CONCEPTS

Platelet	Platelets, or thrombocytes, are small, irregularly shaped clear cell fragments (i.e. cells that do not have a nucleus containing DNA), 2-3 µm in diameter, which are derived from fragmentation of precursor megakaryocytes. The average lifespan of a platelet is normally just 5 to 9 days. Platelets are a natural source of growth factors. They circulate in the blood of mammals and are involved in hemostasis, leading to the formation of blood clots.
Diaphragm	The diaphragm is a cervical barrier type of birth control. It is a soft latex or silicone dome with a spring molded into the rim. The spring creates a seal against the walls of the vagina.
Respiratory system	The respiratory system is the biological system of an organism that introduces respiratory gases to the interior and performs gas exchange. In humans and other mammals, the anatomical features of the respiratory system include airways, lungs, and the respiratory muscles. Molecules of oxygen and carbon dioxide are passively exchanged, by diffusion, between the gaseous external environment and the blood.
Diffusion capacity	In respiratory physiology, diffusing capacity is a long established measurement of the lung's ability to transfer gases between the air and the blood. The words themselves are now misleading because they are archaic: neither is diffusion measured nor is the value obtained from this test a capacity nor even a capacitance, but in fact a conductance. While the term diffusion capacity is retained in the United States for reasons of historical continuity, terminology using transfer factor is now preferred in Europe and elsewhere.
Partial pressure	In a mixture of ideal gases, each gas has a partial pressure which is the pressure which the gas would have if it alone occupied the volume. The total pressure of a gas mixture is the sum of the partial pressures of each individual gas in the mixture.
Prostaglandin	A prostaglandin is any member of a group of lipid compounds that are derived enzymatically from fatty acids and have important functions in the animal body. Every prostaglandin contains 20 carbon atoms, including a 5-carbon ring. They are mediators and have a variety of strong physiological effects, such as regulating the contraction and relaxation of smooth muscle tissue. Prostaglandins are not hormones, but autocrine or paracrine, which are locally acting messenger molecules. They differ from hormones in that they are not produced at a discrete site but in many places throughout the human body. Also, their target cells are present in the immediate vicinity of the site of their secretion (of which there are many).
Respiratory alkalosis	Respiratory alkalosis is a medical condition in which increased respiration (hyperventilation) elevates the blood pH (a condition generally called alkalosis). It is one of four basic categories of disruption of acid-base homeostasis. •Alkalosis refers to a high pH in tissue.•Alkalemia refers to a high pH in the blood.Types

Chapter 11. Physiological adaptation to pregnancy

CHAPTER HIGHLIGHTS & NOTES: KEY TERMS, PEOPLE, PLACES, CONCEPTS

Respiratory rate	Respiratory rate is also known by respiration rate, pulmonary ventilation rate, ventilation rate, or breathing frequency is the number of breaths taken within a set amount of time, typically 60 seconds. A normal respiratory rate is termed eupnea, an increased respiratory rate is termed tachypnea and a lower than normal respiratory rate is termed bradypnea. Human respiration rate is measured when a person is at rest and involves counting the number of breaths for one minute by counting how many times the chest rises.
Oxygen	Oxygen is the element with atomic number 8 and represented by the symbol O. At standard temperature and pressure, two atoms of the element bind to form dioxygen, a colorless, odorless, tasteless diatomic gas with the formula O_2. Oxygen is a member of the chalcogen group on the periodic table, and is a highly reactive nonmetallic element that readily forms compounds (notably oxides) with almost all other elements. By mass, oxygen is the third most abundant element in the universe after hydrogen and helium and the most abundant element by mass in the Earth's crust, making up almost half of the crust's mass.
Ureter	In human anatomy, the ureters are muscular tubes that propel urine from the kidneys to the urinary bladder. In the adult, the ureters are usually 25-30 cm (10-12 in) long and ~3-4 mm in diameter. In humans, the ureters arise from the renal pelvis on the medial aspect of each kidney before descending towards the bladder on the front of the psoas major muscle. The ureters cross the pelvic brim near the bifurcation of the iliac arteries (which they run over). This is a common site for the impaction of kidney stones (the others being the ureterovesical valve and the pelviureteric junction where the ureter joins the renal pelvis in the renal hilum). The ureters run posteroinferiorly on the lateral walls of the pelvis and then curve anteriormedially to enter the bladder through the back, at the vesicoureteric junction, running within the wall of the bladder for a few centimetres. The backflow of urine is prevented by valves known as ureterovesical valves.
Urinary incontinence	Urinary incontinence is any involuntary leakage of urine. It can be a common and distressing problem, which may have a profound impact on quality of life. Urinary incontinence almost always results from an underlying treatable medical condition but is under-reported to medical practitioners.
Urinary system	The urinary system is the organ system that produces, stores, and eliminates urine. In humans it includes two kidneys, two ureters, the bladder, the urethra, and two sphincter muscles.
Urinary tract infection	A urinary tract infection is a bacterial infection that affects part of the urinary tract.

Chapter 11. Physiological adaptation to pregnancy

CHAPTER HIGHLIGHTS & NOTES: KEY TERMS, PEOPLE, PLACES, CONCEPTS

	When it affects the lower urinary tract it is known as a simple cystitis (a bladder infection) and when it affects the upper urinary tract it is known as pyelonephritis (a kidney infection). Symptoms from a lower urinary tract include painful urination and either frequent urination or urge to urinate, while those of pyelonephritis include fever and flank pain in addition to the symptoms of a lower UTI. In the elderly and the very young, symptoms may be vague or non specific.
Progesterone	Progesterone also known as P4 (pregn-4-ene-3,20-dione) is a C-21 steroid hormone involved in the female menstrual cycle, pregnancy (supports gestation) and embryogenesis of humans and other species. Progesterone belongs to a class of hormones called progestogens, and is the major naturally occurring human progestogen. Progesterone was independently discovered by four research groups.
Food craving	A food craving is an intense desire to consume a specific food, stronger than simply normal hunger. According to Marcia Levin Pelchat 'It may be the way in which foods are consumed (e.g. alternating access and restriction) rather than their sensory properties that leads to an addictive eating pattern.' There is no single explanation for food cravings, and explanations range from low serotonin levels affecting the brain centers for appetite to production of endorphins as a result of consuming fats and carbohydrates. Foods with high levels of sugar glucose, such as chocolate, are more frequently craved than foods with lower sugar glucose, such as broccoli because when glucose interacts with opioid system in the brain an addictive triggering effect occurs.
Hyperemesis gravidarum	Hyperemesis gravidarum is a severe form of morning sickness, with 'unrelenting, excessive pregnancy-related nausea and/or vomiting that prevents adequate intake of food and fluids.' Hyperemesis is considered a rare complication of pregnancy but, because nausea and vomiting during pregnancy exist on a continuum, there is often not a good diagnosis between common morning sickness and hyperemesis. Estimates of the percentage of pregnant women afflicted range from 0.3% to 2.0%. Hyperemesis gravidarum is from the Greek hyper-, meaning excessive, and emesis, meaning vomiting, as well as the Latin gravida, meaning pregnant.
Nausea	Nausea, is a sensation of unease and discomfort in the upper stomach with an involuntary urge to vomit. An attack of nausea is known as a qualm.

Chapter 11. Physiological adaptation to pregnancy

CHAPTER HIGHLIGHTS & NOTES: KEY TERMS, PEOPLE, PLACES, CONCEPTS

Vomiting	Vomiting is the forceful expulsion of the contents of one's stomach through the mouth and sometimes the nose. Vomiting can be caused by a wide variety of conditions; it may present as a specific response to ailments like gastritis or poisoning, or as a non-specific sequela of disorders ranging from brain tumors and elevated intracranial pressure to overexposure to ionizing radiation. The feeling that one is about to vomit is called nausea, which usually precedes, but does not always lead to, vomiting.
Large intestine	The large intestine is the third-to-last part of the digestive system in vertebrate animals.
Small intestine	The small intestine is the part of the gastrointestinal tract following the [stomach] and followed by the large intestine, and is where much of the digestion and absorption of food takes place.
Calcium metabolism	Calcium metabolism is the mechanism by which the body maintains adequate calcium levels. Derangements of this mechanism lead to hypercalcemia or hypocalcemia, both of which can have important consequences for health. Calcium is the most abundant mineral in the human body.
Vitamin D	Vitamin D is a group of fat-soluble secosteroids, the two major physiologically relevant forms of which are vitamin D_2 (ergocalciferol) and vitamin D_3 (cholecalciferol). Vitamin D without a subscript refers to either D_2 or D_3 or both. Vitamin D_3 is produced in the skin of vertebrates after exposure to ultraviolet B light from the sun or artificial sources, and occurs naturally in a small range of foods.
Metabolism	Metabolism is the set of chemical reactions that happen in the cells of living organisms to sustain life. These processes allow organisms to grow and reproduce, maintain their structures, and respond to their environments. The word metabolism can also refer to all chemical reactions that occur in living organisms, including digestion and the transport of substances into and between different cells, in which case the set of reactions within the cells is called intermediary metabolism or intermediate metabolism.
Skeleton	The skeleton is the body part that forms the supporting structure of an organism. There are two different skeletal types: the exoskeleton, which is the stable outer shell of an organism, and the endoskeleton, which forms the support structure inside the body. In a figurative sense, skeleton can refer to technology that supports a structure such as a building.
Larynx	The larynx, commonly called the voice box, is an organ in the neck of amphibians, reptiles, birds and mammals (including humans) involved in breathing, sound production, and protecting the trachea against food aspiration. It manipulates pitch and volume.

Chapter 11. Physiological adaptation to pregnancy

CHAPTER HIGHLIGHTS & NOTES: KEY TERMS, PEOPLE, PLACES, CONCEPTS

Diabetes mellitus	Diabetes mellitus, often simply referred to as diabetes, is a group of metabolic diseases in which a person has high blood sugar, either because the body does not produce enough insulin, or because cells do not respond to the insulin that is produced. This high blood sugar produces the classical symptoms of polyuria (frequent urination), polydipsia (increased thirst) and polyphagia (increased hunger). The three main types of diabetes mellitus are:•Type 1 DM results from the body's failure to produce insulin, and presently requires the person to inject insulin.
Gestational diabetes	Gestational diabetes is a condition in which women without previously diagnosed diabetes exhibit high blood glucose levels during pregnancy (especially during third trimester). There is some question whether the condition is natural during pregnancy. Gestational diabetes is caused when the body of a pregnant woman does not secrete enough insulin required during pregnancy, leading to increased blood sugar levels.

CHAPTER QUIZ: KEY TERMS, PEOPLE, PLACES, CONCEPTS

1. _____ is how much three-dimensional space a substance (solid, liquid, gas, or plasma) or shape occupies or contains, often quantified numerically using the SI derived unit, the cubic metre. The _____ of a container is generally understood to be the capacity of the container, i. e. the amount of fluid (gas or liquid) that the container could hold, rather than the amount of space the container itself displaces.

    a. magnetic resonance imaging
    b. Blood sugar regulation
    c. Bovine somatotropin
    d. Volume

2. . Glycoprotein hormone

    _____s are protein hormones secreted by gonadotrope cells of the pituitary gland of vertebrates. This is a family of proteins, which include the mammalian hormones follitropin (FSH), lutropin (LH), placental chorionic _____s hCG and eCG and chorionic _____ as well as at least two forms of fish _____s. These hormones are central to the complex endocrine system that regulates normal growth, sexual development, and reproductive function.

    a. Gonadotropin
    b. Hyperactivation
    c. Hyperestrogenism

Chapter 11. Physiological adaptation to pregnancy

CHAPTER QUIZ: KEY TERMS, PEOPLE, PLACES, CONCEPTS

3. _____ is the element with atomic number 8 and represented by the symbol O. At standard temperature and pressure, two atoms of the element bind to form dioxygen, a colorless, odorless, tasteless diatomic gas with the formula O_2.

_____ is a member of the chalcogen group on the periodic table, and is a highly reactive nonmetallic element that readily forms compounds (notably oxides) with almost all other elements. By mass, _____ is the third most abundant element in the universe after hydrogen and helium and the most abundant element by mass in the Earth's crust, making up almost half of the crust's mass.

a. Izonsteride
b. Oxygen
c. Tracheal intubation
d. SensorMedics High Frequency Oscillatory Ventilator

4. A _____ is a steroid that acts as a hormone. _____ s can be grouped into five groups by the receptors to which they bind: glucocorticoids, mineralocorticoids, androgens, estrogens, and progestogens. Vitamin D derivatives are a sixth closely related hormone system with homologous receptors, though technically sterols rather than steroids.

a. 11-Deoxycorticosterone
b. 17-Hydroxyprogesterone caproate
c. Steroid hormone
d. Pregnenolone

5. _____, $C_{15}H_{12}I_3NO_4$, also known as T_3, is a thyroid hormone. It affects almost every physiological process in the body, including growth and development, metabolism, body temperature, and heart rate.

a. Triiodothyronine
b. Thyroxine
c. Laurel's Kitchen
d. Thyroid acropachy

ANSWER KEY
Chapter 11. Physiological adaptation to pregnancy

1. d
2. a
3. b
4. c
5. a

You can take the complete Chapter Practice Test

for Chapter 11. Physiological adaptation to pregnancy
on all key terms, persons, places, and concepts.

Online 99 Cents

http://www.epub5042.32.20612.11.cram101.com/

Use www.Cram101.com for all your study needs

including Cram101's online interactive problem solving labs in

chemistry, statistics, mathematics, and more.

Chapter 12. Maternal nutrition and health

CHAPTER OUTLINE: KEY TERMS, PEOPLE, PLACES, CONCEPTS

_____ Amino acid

_____ Carbohydrate

_____ Protein

_____ Maternal nutrition

_____ Cholesterol

_____ Fatty acid

_____ High-density lipoprotein

_____ Linoleic acid

_____ Low-density lipoprotein

_____ Toxoplasmosis

_____ Body mass index

_____ Diabetes mellitus

_____ Obesity

_____ Polycystic ovary syndrome

_____ Weight loss

_____ Fetus

_____ Immune system

_____ Albumin

_____ Arachidonic acid

Chapter 12. Maternal nutrition and health

CHAPTER OUTLINE: KEY TERMS, PEOPLE, PLACES, CONCEPTS

| Folic acid |
| Neural tube defect |
| Vital capacity |
| Vitamin D |
| Vitamin K |
| Magnesium |
| Thyroxine |
| Triiodothyronine |
| Malnutrition |
| Fetal alcohol syndrome |

CHAPTER HIGHLIGHTS & NOTES: KEY TERMS, PEOPLE, PLACES, CONCEPTS

Amino acid	Amino acids are molecules containing an amine group, a carboxylic acid group and a side-chain that varies between different amino acids. The key elements of an amino acid are carbon, hydrogen, oxygen, and nitrogen. They are particularly important in biochemistry, where the term usually refers to alpha-amino acids.
Carbohydrate	A carbohydrate is an organic compound that consists only of carbon, hydrogen, and oxygen, usually with a hydrogen:oxygen atom ratio of 2:1 (as in water); in other words, with the empirical formula $C_m(H_2O)_n$. (Some exceptions exist; for example, deoxyribose, a component of DNA, has the empirical formula $C_5H_{10}O_4$). Carbohydrates are not technically hydrates of carbon.
Protein	Proteins are biochemical compounds consisting of one or more polypeptides typically folded into a globular or fibrous form, facilitating a biological function.

Chapter 12. Maternal nutrition and health

CHAPTER HIGHLIGHTS & NOTES: KEY TERMS, PEOPLE, PLACES, CONCEPTS

	A polypeptide is a single linear polymer chain of amino acids bonded together by peptide bonds between the carboxyl and amino groups of adjacent amino acid residues. The sequence of amino acids in a protein is defined by the sequence of a gene, which is encoded in the genetic code.
Maternal nutrition	Maternal nutrition is the dietary intake and habits of expectant mothers with dual emphasis on the health of the mother and the physical and mental development of infants. Nearly 24 per cent of babies are estimated to be born with lower than optimal weights at birth. Pregnancy and child birth transform every aspect of expecting mother's lives. They should be very careful with their choices regarding what they eat.
Cholesterol	Cholesterol is an organic chemical substance classified as a waxy steroid of fat. It is an essential structural component of mammalian cell membranes and is required to establish proper membrane permeability and fluidity. In addition to its importance within cells, cholesterol is an important component in the hormonal systems of the body for the manufacture of bile acids, steroid hormones, and vitamin D. Cholesterol is the principal sterol synthesized by animals; in vertebrates it is formed predominantly in the liver.
Fatty acid	In chemistry, especially biochemistry, a fatty acid is a carboxylic acid with a long aliphatic tail (chain), which is either saturated or unsaturated. Most naturally occurring fatty acids have a chain of an even number of carbon atoms, from 4 to 28. Fatty acids are usually derived from triglycerides or phospholipids. When they are not attached to other molecules, they are known as 'free' fatty acids.
High-density lipoprotein	High-density lipoprotein is one of the five major groups of lipoproteins, which, in order of sizes, largest to smallest, are chylomicrons, VLDL, IDL, LDL, and HDL, which enable lipids like cholesterol and triglycerides to be transported within the water-based bloodstream. In healthy individuals, about thirty percent of blood cholesterol is carried by HDL. Blood tests typically report HDL-C level, i.e. the amount of cholesterol contained in HDL particles. It is often contrasted with low-density or LDL cholesterol or LDL-C. HDL particles are able to remove cholesterol from within artery atheroma and transport it back to the liver for excretion or re-utilization, which is the main reason why the cholesterol carried within HDL particles (HDL-C) is sometimes called 'good cholesterol' (despite the fact that it is exactly the same as the cholesterol in LDL particles).
Linoleic acid	Linoleic acid is an unsaturated n-6 fatty acid. It is a colorless liquid at room temperature.

Chapter 12. Maternal nutrition and health

CHAPTER HIGHLIGHTS & NOTES: KEY TERMS, PEOPLE, PLACES, CONCEPTS

Low-density lipoprotein	Low-density lipoprotein is one of the five major groups of lipoproteins, which in order of size, largest to smallest, are chylomicrons, VLDL, IDL, LDL, and HDL, that enable transport of multiple different fat molecules, including cholesterol, within the water around cells and within the water-based bloodstream. Studies have shown that higher levels of type-B LDL particles (as opposed to type-A LDL particles) promote health problems and cardiovascular disease, they are often informally called the bad cholesterol particles, (as opposed to HDL particles, which are frequently referred to as good cholesterol or healthy cholesterol particles).
	Blood tests typically report LDL-C, the amount of cholesterol contained in LDL. In clinical context, mathematically calculated estimates of LDL-C are commonly used to estimate how much low density lipoproteins are driving progression of atherosclerosis.
Toxoplasmosis	Toxoplasmosis is a parasitic disease caused by the protozoan Toxoplasma gondii. The parasite infects most genera of warm-blooded animals, including humans, but the primary host is the felid (cat) family. Animals are infected by eating infected meat, by ingestion of feces of a cat that has itself recently been infected, or by transmission from mother to fetus.
Body mass index	The body mass index or Quetelet index, is a heuristic proxy for human body fat based on an individual's weight and height. BMI does not actually measure the percentage of body fat. It was devised between 1830 and 1850 by the Belgian polymath Adolphe Quetelet during the course of developing 'social physics'.
Diabetes mellitus	Diabetes mellitus, often simply referred to as diabetes, is a group of metabolic diseases in which a person has high blood sugar, either because the body does not produce enough insulin, or because cells do not respond to the insulin that is produced. This high blood sugar produces the classical symptoms of polyuria (frequent urination), polydipsia (increased thirst) and polyphagia (increased hunger).
	The three main types of diabetes mellitus are:•Type 1 DM results from the body's failure to produce insulin, and presently requires the person to inject insulin.
Obesity	Obesity is a medical condition in which excess body fat has accumulated to the extent that it may have an adverse effect on health, leading to reduced life expectancy and/or increased health problems. Body mass index (BMI), a measurement which compares weight and height, defines people as overweight (pre-obese) if their BMI is between 25 and 30 kg/m^2, and obese when it is greater than 30 kg/m^2.
	Obesity increases the likelihood of various diseases, particularly heart disease, type 2 diabetes, obstructive sleep apnea, certain types of cancer, and osteoarthritis.

Chapter 12. Maternal nutrition and health

CHAPTER HIGHLIGHTS & NOTES: KEY TERMS, PEOPLE, PLACES, CONCEPTS

Polycystic ovary syndrome	Polycystic ovary syndrome is one of the most common female endocrine disorders. PCOS is a complex, heterogeneous disorder of uncertain etiology, but there is strong evidence that it can to a large degree be classified as a genetic disease. PCOS produces symptoms in approximately 5% to 10% of women of reproductive age (12-45 years old).
Weight loss	Weight loss, in the context of medicine, health or physical fitness, is a reduction of the total body mass, due to a mean loss of fluid, body fat or adipose tissue and/or lean mass, namely bone mineral deposits, muscle, tendon and other connective tissue. It can occur unintentionally due to an underlying disease or can arise from a conscious effort to improve an actual or perceived overweight or obese state. Unintentional weight loss Unintentional weight loss occurs in many diseases and conditions, including some very serious diseases such as cancer, AIDS, and a variety of other diseases.
Fetus	A fetus (sometimes spelled foetus or fœtus) is a stage in the development of viviparous organisms. This stage lies between the embryonic stage and birth. The fetuses of most mammals are situated similarly to the homo sapiens fetus within their mothers.
Immune system	The immune system is a system of biological structures and processes within an organism that protects against disease. To function properly, an immune system must detect a wide variety of agents, from viruses to parasitic worms, and distinguish them from the organism's own healthy tissue. Pathogens can rapidly evolve and adapt to avoid detection and neutralization by the immune system.
Albumin	Serum albumin family Albumin refers generally to any protein that is water soluble, is moderately soluble in concentrated salt solutions, and experiences heat denaturation. Albumins are commonly found in blood plasma, and are unique from other blood proteins in that they are not glycosylated. Substances containing albumin, such as egg white, are called albuminoids.
Arachidonic acid	Arachidonic acid is a polyunsaturated omega-6 fatty acid $20:4(\omega-6)$. It is the counterpart to the saturated arachidic acid found in peanut oil, (L. arachis - peanut).

Chapter 12. Maternal nutrition and health

CHAPTER HIGHLIGHTS & NOTES: KEY TERMS, PEOPLE, PLACES, CONCEPTS

	Arachidonic acid is one of the essential fatty acids required by most mammals. Some mammals lack the ability to--or have a very limited capacity to--convert linoleic acid into arachidonic acid, making it an essential part of their diet. Since little or no arachidonic acid is found in common plants, such animals are obligate carnivores; the cat is a common example. A commercial source of arachidonic acid has been derived, however, from the fungus Mortierella alpina
Folic acid	Folic acid and folate (the form naturally occurring in the body), as well as pteroyl-L-glutamic acid, pteroyl-L-glutamate, and pteroylmonoglutamic acid are forms of the water-soluble vitamin B_9. Folic acid is itself not biologically active, but its biological importance is due to tetrahydrofolate and other derivatives after its conversion to dihydrofolic acid in the liver. Vitamin B_9 (folic acid and folate inclusive) is essential to numerous bodily functions.
Neural tube defect	Neural tube defects are one of the most common birth defects, occurring in approximately one in 1,000 live births in the United States. An Neural tube defect is an opening in the spinal cord or brain that occurs very early in human development. In about the 3rd or 4th weeks of pregnancy, specialized cells on the dorsal side of the fetus begin to fuse and form the neural tube.
Vital capacity	Vital capacity is the maximum amount of air a person can expel from the lungs after a maximum inhalation. It is equal to the inspiratory reserve volume plus the tidal volume plus the expiratory reserve volume. A person's vital capacity can be measured by a spirometer which can be a wet or regular spirometer.
Vitamin D	Vitamin D is a group of fat-soluble secosteroids, the two major physiologically relevant forms of which are vitamin D_2 (ergocalciferol) and vitamin D_3 (cholecalciferol). Vitamin D without a subscript refers to either D_2 or D_3 or both. Vitamin D_3 is produced in the skin of vertebrates after exposure to ultraviolet B light from the sun or artificial sources, and occurs naturally in a small range of foods.
Vitamin K	Vitamin K is a group of structurally similar, fat-soluble vitamins that are needed for the posttranslational modification of certain proteins required for blood coagulation and in metabolic pathways in bone and other tissue. They are 2-methyl-1,4-naphthoquinone (3-) derivatives. This group of vitamins includes two natural vitamers: vitamin K_1 and vitamin K_2.
Magnesium	Magnesium is a chemical element with the symbol Mg, atomic number 12, and common oxidation number +2. It is an alkaline earth metal and the eighth most abundant element in the Earth's crust, where it constitutes about 2% by mass, and ninth in the known universe as a whole.

Chapter 12. Maternal nutrition and health

CHAPTER HIGHLIGHTS & NOTES: KEY TERMS, PEOPLE, PLACES, CONCEPTS

	This preponderance of magnesium is related to the fact that it is easily built up in supernova stars from a sequential addition of three helium nuclei to carbon (which in turn is made from three helium nuclei). Due to magnesium ion's high solubility in water, it is the third most abundant element dissolved in seawater.
Thyroxine	Thyroxine, or 3,5,3',5'-tetraiodothyronine, a form of thyroid hormones is the major hormone secreted by the follicular cells of the thyroid gland. Synthesis and regulation Thyroxine is synthesized via the iodination and covalent bonding of the phenyl portions of tyrosine residues found in an initial peptide, thyroglobulin, which is secreted into thyroid granules. These iodinated diphenyl compounds are cleaved from their peptide backbone upon being stimulated by thyroid-stimulating hormone.
Triiodothyronine	Triiodothyronine, $C_{15}H_{12}I_3NO_4$, also known as T_3, is a thyroid hormone. It affects almost every physiological process in the body, including growth and development, metabolism, body temperature, and heart rate.
Malnutrition	Malnutrition is the condition that results from taking an unbalanced diet in which certain nutrients are lacking, in excess (too high an intake), or in the wrong proportions. A number of different nutrition disorders may arise, depending on which nutrients are under or overabundant in the diet. In most of the world, malnutrition is present in the form of undernutrition, which is caused by a diet lacking adequate calories and protein.
Fetal alcohol syndrome	Fetal alcohol syndrome is a pattern of mental and physical defects that can develop in a fetus in association with high levels of alcohol consumption during pregnancy. Current research also implicates other lifestyle choices made by the prospective mother. Indications for lower levels of alcohol are inconclusive.

Chapter 12. Maternal nutrition and health

CHAPTER QUIZ: KEY TERMS, PEOPLE, PLACES, CONCEPTS

1. _____ is the condition that results from taking an unbalanced diet in which certain nutrients are lacking, in excess (too high an intake), or in the wrong proportions. A number of different nutrition disorders may arise, depending on which nutrients are under or overabundant in the diet. In most of the world, _____ is present in the form of undernutrition, which is caused by a diet lacking adequate calories and protein.

   a. Malnutrition-inflammation complex
   b. Medical food
   c. Metabolic advantage
   d. Malnutrition

2. _____, often simply referred to as diabetes, is a group of metabolic diseases in which a person has high blood sugar, either because the body does not produce enough insulin, or because cells do not respond to the insulin that is produced. This high blood sugar produces the classical symptoms of polyuria (frequent urination), polydipsia (increased thirst) and polyphagia (increased hunger).

   The three main types of _____ are:•Type 1 DM results from the body's failure to produce insulin, and presently requires the person to inject insulin.

   a. Dietary Reference Values
   b. Health claims on food labels
   c. Diabetes mellitus
   d. History of USDA nutrition guides

3. In chemistry, especially biochemistry, a _____ is a carboxylic acid with a long aliphatic tail (chain), which is either saturated or unsaturated. Most naturally occurring _____s have a chain of an even number of carbon atoms, from 4 to 28. _____s are usually derived from triglycerides or phospholipids. When they are not attached to other molecules, they are known as 'free' _____s.

   a. Fatty acid
   b. Flavonoid
   c. Fluid balance
   d. Food Balance Wheel

4. _____s are molecules containing an amine group, a carboxylic acid group and a side-chain that varies between different _____s. The key elements of an _____ are carbon, hydrogen, oxygen, and nitrogen. They are particularly important in biochemistry, where the term usually refers to alpha-_____s.

   a. Amino acid
   b. Alosetron
   c. Adhesion
   d. Algaemia

5. . A _____(sometimes spelled foetus or fœtus) is a stage in the development of viviparous organisms.

Chapter 12. Maternal nutrition and health

CHAPTER QUIZ: KEY TERMS, PEOPLE, PLACES, CONCEPTS

This stage lies between the embryonic stage and birth.

The fetuses of most mammals are situated similarly to the homo sapiens fetus within their mothers.

a. Gestation period
b. Fetus
c. Habitual abortion
d. Human fertilization

ANSWER KEY
Chapter 12. Maternal nutrition and health

1. d
2. c
3. a
4. a
5. b

You can take the complete Chapter Practice Test

for Chapter 12. Maternal nutrition and health
on all key terms, persons, places, and concepts.

Online 99 Cents

http://www.epub5042.32.20612.12.cram101.com/

Use www.Cram101.com for all your study needs

including Cram101's online interactive problem solving labs in

chemistry, statistics, mathematics, and more.

Chapter 13. Physiology of parturition

CHAPTER OUTLINE: KEY TERMS, PEOPLE, PLACES, CONCEPTS

_____ Uterine contraction

_____ Caesarean section

_____ Adipose tissue

_____ Fetus

_____ Immune system

_____ Actin

_____ Calmodulin

_____ Myometrium

_____ Braxton Hicks contractions

_____ Fibronectin

_____ Effacement

_____ Adrenocorticotropic hormone

_____ Animal model

_____ Cortisol

_____ Hypertension

_____ Corticosteroid

_____ Mifepristone

_____ Receptor

_____ Arachidonic acid

Chapter 13. Physiology of parturition
CHAPTER OUTLINE: KEY TERMS, PEOPLE, PLACES, CONCEPTS

	Prostaglandin
	Oxytocin
	Prostacyclin
	Nervous system
	Syndrome
	Pelvic floor
	Kidney
	Renin-angiotensin system
	Respiratory alkalosis
	Respiratory system
	Nutrition
	Ketone bodies
	Ketosis
	Asphyxia
	Heart rate
	Meconium
	Metabolic acidosis
	Presentation
	Swallowing

Chapter 13. Physiology of parturition

CHAPTER OUTLINE: KEY TERMS, PEOPLE, PLACES, CONCEPTS

	Blood flow
	Lung
	Transmission
	Visceral pain
	Muscle contraction
	Somatic

CHAPTER HIGHLIGHTS & NOTES: KEY TERMS, PEOPLE, PLACES, CONCEPTS

Uterine contraction	A uterine contraction is a muscle contraction of the uterine smooth muscle. The uterus frequently contracts throughout the entire menstrual cycle, and these contractions have been termed endometrial waves or contractile waves. These appear to involve only the sub-endometrial layer of the myometrium.
Caesarean section	A Caesarean section, (also C-section, Caesarian section, Cesarean section, Caesar, etc). is a surgical procedure in which one or more incisions are made through a mother's abdomen (laparotomy) and uterus (hysterotomy) to deliver one or more babies, or, rarely, to remove a dead fetus. A late-term abortion using Caesarean section procedures is termed a hysterotomy abortion and is very rarely performed.
Adipose tissue	In histology, adipose tissue is loose connective tissue composed of adipocytes. It is technically composed of roughly only 80% fat; fat in its solitary state exists in the liver and muscles. Adipose tissue is derived from lipoblasts.
Fetus	A fetus (sometimes spelled foetus or fœtus) is a stage in the development of viviparous organisms. This stage lies between the embryonic stage and birth.

Chapter 13. Physiology of parturition

CHAPTER HIGHLIGHTS & NOTES: KEY TERMS, PEOPLE, PLACES, CONCEPTS

Immune system	The immune system is a system of biological structures and processes within an organism that protects against disease. To function properly, an immune system must detect a wide variety of agents, from viruses to parasitic worms, and distinguish them from the organism's own healthy tissue.
	Pathogens can rapidly evolve and adapt to avoid detection and neutralization by the immune system.
Actin	Actin is a globular, roughly 42-kDa protein found in all eukaryotic cells (the only known exception being nematode sperm) where it may be present at concentrations of over 100 μM. It is also one of the most highly-conserved proteins, differing by no more than 20% in species as diverse as algae and humans. Actin is the monomeric subunit of two types of filaments in cells: microfilaments, one of the three major components of the cytoskeleton, and thin filaments, part of the contractile apparatus in muscle cells. Thus, actin participates in many important cellular processes including muscle contraction, cell motility, cell division and cytokinesis, vesicle and organelle movement, cell signaling, and the establishment and maintenance of cell junctions and cell shape.
Calmodulin	Calmodulin (an abbreviation for CALcium-MODULated proteIN) is a calcium-binding messenger protein expressed in all eukaryotic cells. CaM is a multifunctional intermediate messenger protein that transduces calcium signals by binding calcium ions and then modifying its interactions with various target proteins.
	CaM mediates many crucial processes such as inflammation, metabolism, apoptosis, smooth muscle contraction, intracellular movement, short-term and long-term memory, and the immune response.
Myometrium	The myometrium is the middle layer of the uterine wall, consisting mainly of uterine smooth muscle cells (also called uterine myocytes), but also of supporting stromal and vascular tissue. Its main function is to induce uterine contractions.
	The myometrium is located between the endometrium (the inner layer of the uterine wall), and the serosa or perimetrium (the outer uterine layer).
Braxton Hicks contractions	Braxton Hicks contractions, are sporadic uterine contractions that sometimes start around 6 weeks. However, they are not usually felt until the second trimester or third trimester of pregnancy.
	They should be infrequent, irregular, and involve only mild cramping.

Chapter 13. Physiology of parturition

CHAPTER HIGHLIGHTS & NOTES: KEY TERMS, PEOPLE, PLACES, CONCEPTS

Fibronectin	Fibronectin is a high-molecular weight (~440kDa) extracellular matrix glycoprotein that binds to membrane-spanning receptor proteins called integrins. In addition to integrins, fibronectin also binds extracellular matrix components such as collagen, fibrin and heparan sulfate proteoglycans (e.g. syndecans).
	Fibronectin exists as a dimer, consisting of two nearly identical monomers linked by a pair of disulfide bonds. The fibronectin protein is produced from a single gene, but alternative splicing of its pre-mRNA leads to the creation of several isoforms.
Effacement	Effacement is the shortening, or thinning, of a tissue.
	It can refer to cervical effacement. It can also refer to a process occurring in podocytes.
Adrenocorticotropic hormone	Adrenocorticotropic hormone also known as corticotropin, is a polypeptide tropic hormone produced and secreted by the anterior pituitary gland. It is an important component of the hypothalamic-pituitary-adrenal axis and is often produced in response to biological stress (along with its precursor corticotropin-releasing hormone from the hypothalamus). Its principal effects are increased production and release of corticosteroids.
Animal model	An animal model is a living, non-human animal used during the research and investigation of human disease, for the purpose of better understanding the disease without the added risk of causing harm to an actual human being during the process. The animal chosen will usually meet a determined taxonomic equivalency to humans, so as to react to disease or its treatment in a way that resembles human physiology as needed. Many drugs, treatments and cures for human diseases have been developed with the use of animal models.
Cortisol	Cortisol is a steroid hormone, or glucocorticoid, produced by the adrenal gland. It is released in response to stress and a low level of blood glucocorticoids. Its primary functions are to increase blood sugar through gluconeogenesis; suppress the immune system; and aid in fat, protein and carbohydrate metabolism.
Hypertension	Hypertension or high blood pressure, sometimes called arterial hypertension, is a chronic medical condition in which the blood pressure in the arteries is elevated. This requires the heart to work harder than normal to circulate blood through the blood vessels. Blood pressure involves two measurements, systolic and diastolic, which depend on whether the heart muscle is contracting (systole) or relaxed between beats (diastole).
Corticosteroid	Corticosteroids are a class of chemicals that includes steroid hormones naturally produced in the adrenal cortex of vertebrates and analogues of these hormones that are synthesized in laboratories.

Chapter 13. Physiology of parturition

CHAPTER HIGHLIGHTS & NOTES: KEY TERMS, PEOPLE, PLACES, CONCEPTS

	Corticosteroids are involved in a wide range of physiologic processes, including stress response, immune response, and regulation of inflammation, carbohydrate metabolism, protein catabolism, blood electrolyte levels, and behavior. •Glucocorticoids such as cortisol control carbohydrate, fat and protein metabolism and are anti-inflammatory by preventing phospholipid release, decreasing eosinophil action and a number of other mechanisms.•Mineralocorticoids such as aldosterone control electrolyte and water levels, mainly by promoting sodium retention in the kidney.
	Some common natural hormones are corticosterone ($C_{21}H_{30}O_4$), cortisone ($C_{21}H_{28}O_5$, 17-hydroxy-11-dehydrocorticosterone) and aldosterone.
Mifepristone	Mifepristone is a synthetic steroid compound used as a pharmaceutical. It is a progesterone receptor antagonist used as an abortifacient in the first months of pregnancy, and in smaller doses as an emergency contraceptive. Mifepristone is also a powerful glucocorticoid receptor antagonist, and has occasionally been used in refractory Cushing's Syndrome (due to ectopic/neoplastic ACTH/Cortisol secretion).
Receptor	In biochemistry, a receptor is a protein molecule, embedded in either the plasma membrane or the cytoplasm of a cell, to which one or more specific kinds of signaling molecules may attach. A molecule which binds (attaches) to a receptor is called a ligand, and may be a peptide (short protein) or other small molecule, such as a neurotransmitter, a hormone, a pharmaceutical drug, or a toxin. Each kind of receptor can bind only certain ligand shapes. Each cell typically has many receptors, of many different kinds. Simply put, a receptor functions as a keyhole that opens a neural path when the proper ligand is inserted.
Arachidonic acid	Arachidonic acid is a polyunsaturated omega-6 fatty acid 20:4(ω-6). It is the counterpart to the saturated arachidic acid found in peanut oil, (L. arachis - peanut).
	Arachidonic acid is one of the essential fatty acids required by most mammals. Some mammals lack the ability to--or have a very limited capacity to--convert linoleic acid into arachidonic acid, making it an essential part of their diet. Since little or no arachidonic acid is found in common plants, such animals are obligate carnivores; the cat is a common example. A commercial source of arachidonic acid has been derived, however, from the fungus Mortierella alpina
Prostaglandin	A prostaglandin is any member of a group of lipid compounds that are derived enzymatically from fatty acids and have important functions in the animal body. Every prostaglandin contains 20 carbon atoms, including a 5-carbon ring.
	They are mediators and have a variety of strong physiological effects, such as regulating the contraction and relaxation of smooth muscle tissue. Prostaglandins are not hormones, but autocrine or paracrine, which are locally acting messenger molecules.

Chapter 13. Physiology of parturition

CHAPTER HIGHLIGHTS & NOTES: KEY TERMS, PEOPLE, PLACES, CONCEPTS

	They differ from hormones in that they are not produced at a discrete site but in many places throughout the human body. Also, their target cells are present in the immediate vicinity of the site of their secretion (of which there are many).
Oxytocin	Oxytocin ( ) is a mammalian hormone that acts primarily as a neuromodulator in the brain. Oxytocin is best known for its roles in sexual reproduction, in particular during and after childbirth. It is released in large amounts after distension of the cervix and uterus during labor, facilitating birth, and after stimulation of the nipples, facilitating breastfeeding.
Prostacyclin	Prostacyclin is a member of the family of lipid molecules known as eicosanoids. As a drug, it is also known as 'epoprostenol'. The terms are sometimes used interchangeably.
Nervous system	The nervous system is an organ system containing a network of specialized cells called neurons that coordinate the actions of an animal and transmit signals between different parts of its body. In most animals the nervous system consists of two parts, central and peripheral. The central nervous system of vertebrates (such as humans) contains the brain, spinal cord, and retina. The peripheral nervous system consists of sensory neurons, clusters of neurons called ganglia, and nerves connecting them to each other and to the central nervous system. These regions are all interconnected by means of complex neural pathways. The enteric nervous system, a subsystem of the peripheral nervous system, has the capacity, even when severed from the rest of the nervous system through its primary connection by the vagus nerve, to function independently in controlling the gastrointestinal system.
Syndrome	The term syndrome derives from the Greek συνδρομή (sundrome) and means 'concurrence of symptoms, concourse', from σύν (sun), 'along with, together' + δρόμος (dromos), amongst others 'course'. In medicine and psychology, a syndrome is the association of several clinically recognizable features, signs (observed by someone other than the patient), symptoms (reported by the patient), phenomena or characteristics that often occur together, so that the presence of one or more features alerts the healthcare provider to the possible presence of the others. In recent decades, the term has been used outside medicine to refer to a combination of phenomena seen in association.
Pelvic floor	The pelvic floor is composed of muscle fibers of the levator ani, the coccygeus, and associated connective tissue which span the area underneath the pelvis. The pelvic diaphragm is a muscular partition formed by the levatores ani and coccygei, with which may be included the parietal pelvic fascia on their upper and lower aspects.

Chapter 13. Physiology of parturition

CHAPTER HIGHLIGHTS & NOTES: KEY TERMS, PEOPLE, PLACES, CONCEPTS

Kidney	The kidneys are organs with several functions. They are seen in many types of animals, including vertebrates and some invertebrates. They are an essential part of the urinary system and also serve homeostatic functions such as the regulation of electrolytes, maintenance of acid-base balance, and regulation of blood pressure. They serve the body as a natural filter of the blood, and remove wastes which are diverted to the urinary bladder. In producing urine, the kidneys excrete wastes such as urea and ammonium; the kidneys also are responsible for the reabsorption of water, glucose, and amino acids. The kidneys also produce hormones including calcitriol, renin, and erythropoietin.
Renin-angiotensin system	The renin-angiotensin system or the renin-angiotensin-aldosterone system (RAAS) is a hormone system that regulates blood pressure and water (fluid) balance. When blood volume is low, juxtaglomerular cells in the kidneys secrete renin. Renin stimulates the production of angiotensin I, which is then converted to angiotensin II. Angiotensin II causes blood vessels to constrict, resulting in increased blood pressure.
Respiratory alkalosis	Respiratory alkalosis is a medical condition in which increased respiration (hyperventilation) elevates the blood pH (a condition generally called alkalosis). It is one of four basic categories of disruption of acid-base homeostasis. •Alkalosis refers to a high pH in tissue.•Alkalemia refers to a high pH in the blood.Types There are two types of respiratory alkalosis: chronic and acute.
Respiratory system	The respiratory system is the biological system of an organism that introduces respiratory gases to the interior and performs gas exchange. In humans and other mammals, the anatomical features of the respiratory system include airways, lungs, and the respiratory muscles. Molecules of oxygen and carbon dioxide are passively exchanged, by diffusion, between the gaseous external environment and the blood.
Nutrition	Nutrition is the provision, to cells and organisms, of the materials necessary (in the form of food) to support life. Many common health problems can be prevented or alleviated with a healthy diet. The diet of an organism is what it eats, which is largely determined by the perceived palatability of foods.
Ketone bodies	Ketone bodies are three water-soluble compounds that are produced as by-products when fatty acids are broken down for energy in the liver. Two of the three are used as a source of energy in the heart and brain while the third is a waste product excreted from the body. In the brain, they are a vital source of energy during fasting.
Ketosis	Ketosis is a state of elevated levels of ketone bodies in the body.

Chapter 13. Physiology of parturition

CHAPTER HIGHLIGHTS & NOTES: KEY TERMS, PEOPLE, PLACES, CONCEPTS

	It is almost always generalized throughout the body, with hyperketonemia, that is, an elevated level of ketone bodies in the blood. Ketone bodies are formed by ketogenesis when the liver glycogen stores are depleted.
Asphyxia	Asphyxia is a condition of severely deficient supply of oxygen to the body that arises from being unable to breathe normally. An example of asphyxia is choking. Asphyxia causes generalized hypoxia, which primarily affects the tissues and organs.
Heart rate	Heart rate is the number of heartbeats per unit of time, typically expressed as beats per minute (bpm). Heart rate can vary as the body's need to absorb oxygen and excrete carbon dioxide changes, such as during exercise or sleep. The measurement of heart rate is used by medical professionals to assist in the diagnosis and tracking of medical conditions.
Meconium	Meconium is the earliest stools of an infant. Unlike later feces, meconium is composed of materials ingested during the time the infant spends in the uterus: intestinal epithelial cells, lanugo, mucus, amniotic fluid, bile, and water. Meconium is almost sterile, unlike later feces, is viscous and sticky like tar, and has no odor.
Metabolic acidosis	In medicine, metabolic acidosis is a condition that occurs when the body produces too much acid or when the kidneys are not removing enough acid from the body. If unchecked, metabolic acidosis leads to acidemia, i.e., blood pH is low (less than 7.35) due to increased production of hydrogen by the body or the inability of the body to form bicarbonate (HCO_3^-) in the kidney. Its causes are diverse, and its consequences can be serious, including coma and death.
Presentation	In obstetrics, the presentation of a fetus about to be born refers to which anatomical part of the fetus is leading, that is, is closest to the pelvic inlet of the birth canal. According to the leading part, this is identified as a cephalic, breech, or shoulder presentation. A malpresentation is any other presentation than a vertex presentation.
Swallowing	Swallowing, known scientifically as deglutition, is the process in the human or animal body that makes something pass from the mouth, to the pharynx, and into the esophagus, while shutting the epiglottis. If this fails and the object goes through the trachea, then choking or pulmonary aspiration can occur. In the human body it is controlled by the swallowing reflex.
Blood flow	Blood flow is the continuous running of blood in the cardiovascular system. The human body is made up of several processes all carrying out various functions. We have the gastrointestinal system which aids the digestion and the absorption of food.

Chapter 13. Physiology of parturition

CHAPTER HIGHLIGHTS & NOTES: KEY TERMS, PEOPLE, PLACES, CONCEPTS

Lung	The lung is the essential respiration cell in many air-breathing animals, including most tetrapods, a few fish and a few snails. In mammals and the more complex life forms, the two lungs are located near the backbone on either side of the heart. Their principal function is to transport oxygen from the atmosphere into the bloodstream, and to release carbon dioxide from the bloodstream into the atmosphere. This exchange of gases is accomplished in the mosaic of specialized cells that form millions of tiny, exceptionally thin-walled air sacs called alveoli.
Transmission	In medicine and biology, transmission is the passing of a communicable disease from an infected host individual or group to a conspecific individual or group, regardless of whether the other individual was previously infected . Sometimes transmission can specifically mean infection of a previously uninfected host. The term usually refers to the transmission of microorganisms directly from one person to another by one or more of the following means:•droplet contact - coughing or sneezing on another person•direct physical contact - touching an infected person, including sexual contact•indirect physical contact - usually by touching soil contamination or a contaminated surface•airborne transmission - if the microorganism can remain in the air for long periods•fecal-oral transmission - usually from contaminated food or water sources Transmission can also be indirect, via another organism, either a vector (e.g. a mosquito) or an intermediate host (e.g. tapeworm in pigs can be transmitted to humans who ingest improperly cooked pork).
Visceral pain	Visceral Pain is pain that results from the activation of nociceptors of the thoracic, pelvic, or abdominal viscera (organs). Visceral structures are highly sensitive to distension (stretch), ischemia and inflammation, but relatively insensitive to other stimuli that normally evoke pain such as cutting or burning. Visceral pain is diffuse, difficult to localize and often referred to a distant, usually superficial, structure.
Muscle contraction	Muscle fiber generates tension through the action of actin and myosin cross-bridge cycling. While under tension, the muscle may lengthen, shorten, or remain the same. Although the term contraction implies shortening, when referring to the muscular system, it means muscle fibers generating tension with the help of motor neurons (the terms twitch tension, twitch force, and fiber contraction are also used). Voluntary muscle contraction is controlled by the central nervous system.
Somatic	The term somatic means 'of the body'. It is often used in biology to refer to the cells of the body in contrast to the cells in the germline which give rise to the gametes (eggs or sperm). These somatic cells are diploid containing two copies of each chromosome, whereas the germ cells are haploid as they only contain one copy of each chromosome.

Chapter 13. Physiology of parturition

CHAPTER HIGHLIGHTS & NOTES: KEY TERMS, PEOPLE, PLACES, CONCEPTS

Although under normal circumstances all somatic cells contain identical DNA, they develop a variety of tissue-specific characteristics. This process is called differentiation, through epigenetic and regulatory alterations. The grouping of like cells and tissues creates the foundation for organs.

CHAPTER QUIZ: KEY TERMS, PEOPLE, PLACES, CONCEPTS

1. In biochemistry, a _____ is a protein molecule, embedded in either the plasma membrane or the cytoplasm of a cell, to which one or more specific kinds of signaling molecules may attach. A molecule which binds (attaches) to a _____ is called a ligand, and may be a peptide (short protein) or other small molecule, such as a neurotransmitter, a hormone, a pharmaceutical drug, or a toxin. Each kind of _____ can bind only certain ligand shapes. Each cell typically has many _____s, of many different kinds. Simply put, a _____ functions as a keyhole that opens a neural path when the proper ligand is inserted.

    a. magnetic resonance imaging
    b. Receptor
    c. Cloprednol
    d. Corticosterone

2. _____ is a medical condition in which increased respiration (hyperventilation) elevates the blood pH (a condition generally called alkalosis). It is one of four basic categories of disruption of acid-base homeostasis. •Alkalosis refers to a high pH in tissue.•Alkalemia refers to a high pH in the blood.Types

    There are two types of _____: chronic and acute.

    a. Respiratory alkalosis
    b. Blood sugar regulation
    c. Bovine somatotropin
    d. Chromatophobe

3. _____ is a member of the family of lipid molecules known as eicosanoids.

    As a drug, it is also known as 'epoprostenol'. The terms are sometimes used interchangeably.

    a. Prulifloxacin
    b. Prostacyclin
    c. Quinic acid
    d. Racemic acid

Chapter 13. Physiology of parturition

CHAPTER QUIZ: KEY TERMS, PEOPLE, PLACES, CONCEPTS

4. The _____s are organs with several functions. They are seen in many types of animals, including vertebrates and some invertebrates. They are an essential part of the urinary system and also serve homeostatic functions such as the regulation of electrolytes, maintenance of acid-base balance, and regulation of blood pressure. They serve the body as a natural filter of the blood, and remove wastes which are diverted to the urinary bladder. In producing urine, the _____s excrete wastes such as urea and ammonium; the _____s also are responsible for the reabsorption of water, glucose, and amino acids. The _____s also produce hormones including calcitriol, renin, and erythropoietin.

   a. magnetic resonance imaging
   b. Spermatheca
   c. Kidney
   d. Vaginal flatulence

5. _____ is the shortening, or thinning, of a tissue.

   It can refer to cervical _____. It can also refer to a process occurring in podocytes.

   a. Effacement
   b. Epithelium
   c. Extracellular matrix
   d. Iliac fascia

ANSWER KEY
Chapter 13. Physiology of parturition

1. b
2. a
3. b
4. c
5. a

You can take the complete Chapter Practice Test

for Chapter 13. Physiology of parturition
on all key terms, persons, places, and concepts.

Online 99 Cents

http://www.epub5042.32.20612.13.cram101.com/

Use www.Cram101.com for all your study needs

including Cram101's online interactive problem solving labs in

chemistry, statistics, mathematics, and more.

Chapter 14. The puerperium

CHAPTER OUTLINE: KEY TERMS, PEOPLE, PLACES, CONCEPTS

	Retained placenta
	Cervix
	Oxytocin
	Product
	Episiotomy
	Soft tissue
	Wound healing
	Lochia
	Curettage
	Deep vein
	Deep vein thrombosis
	Respiratory system
	Urinary system
	Urinary tract infection
	Urinary catheterization
	Vagina
	Pelvic floor
	Libido
	Thrombophlebitis

Chapter 14. The puerperium
CHAPTER OUTLINE: KEY TERMS, PEOPLE, PLACES, CONCEPTS

	Infection
	Pulmonary embolism
	Breastfeeding
	Postnatal

CHAPTER HIGHLIGHTS & NOTES: KEY TERMS, PEOPLE, PLACES, CONCEPTS

Retained placenta	Retained placenta is a condition where all or part of the placenta or membranes are left behind in the uterus during the third stage of labour. In humans, retained placenta is generally defined as a placenta that has not undergone placental expulsion within 30 minutes of the baby's birth. Risks of retained placenta include hemorrhage and infection.
Cervix	The cervix is the lower, narrow portion of the uterus where it joins with the top end of the vagina. It is cylindrical or conical in shape and protrudes through the upper anterior vaginal wall. Approximately half its length is visible with appropriate medical equipment; the remainder lies above the vagina beyond view.
Oxytocin	Oxytocin ( ) is a mammalian hormone that acts primarily as a neuromodulator in the brain. Oxytocin is best known for its roles in sexual reproduction, in particular during and after childbirth. It is released in large amounts after distension of the cervix and uterus during labor, facilitating birth, and after stimulation of the nipples, facilitating breastfeeding.
Product	In biochemistry, a product is something 'manufactured' by an enzyme from its substrate. For example the products of Lactase are Galactose and Glucose, which are produced from the substrate Lactose.
Episiotomy	An episiotomy also known as perineotomy, is a surgically planned incision on the perineum and the posterior vaginal wall during second stage of labor.

Chapter 14. The puerperium

CHAPTER HIGHLIGHTS & NOTES: KEY TERMS, PEOPLE, PLACES, CONCEPTS

	The incision, which can be midline or at an angle from the posterior end of the vulva, is performed under local anesthetic (pudendal anesthesia), and is sutured closed after delivery. It is one of the most common medical procedures performed on women, and although its routine use in childbirth has steadily declined in recent decades, it is still widely practiced in many parts of the world including Latin America, Poland, Bulgaria, India and Taiwan.
Soft tissue	In anatomy, the term soft tissue refers to tissues that connect, support, or surround other structures and organs of the body, not being bone. Soft tissue includes tendons, ligaments, fascia, skin, fibrous tissues, fat, and synovial membranes (which are connective tissue), and muscles, nerves and blood vessels (which are not connective tissue). It is sometimes defined by what it is not.
Wound healing	Wound healing, is an intricate process in which the skin repairs itself after injury. In normal skin, the epidermis (outermost layer) and dermis (inner or deeper layer) exists in a steady-state equilibrium, forming a protective barrier against the external environment. Once the protective barrier is broken, the normal (physiologic) process of wound healing is immediately set in motion.
Lochia	In the field of obstetrics, lochia is post-partum vaginal discharge, containing blood, mucus, and placental tissue. Lochia discharge typically continues for 4 to 6 weeks after childbirth. It progresses through three stages.
Curettage	Curettage in medical procedures, is the use of a curette to remove tissue by scraping or scooping. Curettages are also a declining method of abortion. It has been replaced by vacuum aspiration over the last decade.
Deep vein	Deep vein is a term used to describe a vein that is deep in the body. It is used to differentiate deep veins from veins which are close to the surface, also known as superficial veins. Deep veins are almost always beside an artery with the same name (e.g. the femoral vein is beside the femoral artery).
Deep vein thrombosis	A deep vein thrombosis is a blood clot in a deep vein. A clot inside a blood vessel is called a thrombosis. DVTs predominantly occur in the legs and may have no symptoms.
Respiratory system	The respiratory system is the biological system of an organism that introduces respiratory gases to the interior and performs gas exchange.

Chapter 14. The puerperium

CHAPTER HIGHLIGHTS & NOTES: KEY TERMS, PEOPLE, PLACES, CONCEPTS

	In humans and other mammals, the anatomical features of the respiratory system include airways, lungs, and the respiratory muscles. Molecules of oxygen and carbon dioxide are passively exchanged, by diffusion, between the gaseous external environment and the blood.
Urinary system	The urinary system is the organ system that produces, stores, and eliminates urine. In humans it includes two kidneys, two ureters, the bladder, the urethra, and two sphincter muscles.
Urinary tract infection	A urinary tract infection is a bacterial infection that affects part of the urinary tract. When it affects the lower urinary tract it is known as a simple cystitis (a bladder infection) and when it affects the upper urinary tract it is known as pyelonephritis (a kidney infection). Symptoms from a lower urinary tract include painful urination and either frequent urination or urge to urinate , while those of pyelonephritis include fever and flank pain in addition to the symptoms of a lower UTI. In the elderly and the very young, symptoms may be vague or non specific.
Urinary catheterization	In urinary catheterization a latex, polyurethane, or silicone tube known as a urinary catheter is inserted into a patient's bladder via the urethra. Catheterization allows the patient's urine to drain freely from the bladder for collection. It may be used to inject liquids used for treatment or diagnosis of bladder conditions.
Vagina	The vagina is a fibromuscular tubular tract which is a sex organ and has two main functions; sexual intercourse and childbirth. In humans, this passage leads from the opening of the vulva to the uterus (womb), but the vaginal tract ends at the cervix. Unlike men, who have only one genital orifice, women have two, the urethra and the vagina.
Pelvic floor	The pelvic floor is composed of muscle fibers of the levator ani, the coccygeus, and associated connective tissue which span the area underneath the pelvis. The pelvic diaphragm is a muscular partition formed by the levatores ani and coccygei, with which may be included the parietal pelvic fascia on their upper and lower aspects. The pelvic floor separates the pelvic cavity above from the perineal region (including perineum) below.
Libido	Libido refers to a person's sex drive or desire for sexual activity. The desire for sex is an aspect of a person's sexuality, but varies enormously from one person to another, and it also varies depending on circumstances at a particular time. A person who has extremely frequent or a suddenly increased sex drive may be experiencing hypersexuality.
Thrombophlebitis	Thrombophlebitis is phlebitis (vein inflammation) related to a thrombus (blood clot). When it occurs repeatedly in different locations, it is known as 'Thrombophlebitis migrans' or 'migrating thrombophlebitis'.

Chapter 14. The puerperium

CHAPTER HIGHLIGHTS & NOTES: KEY TERMS, PEOPLE, PLACES, CONCEPTS

	The following symptoms are often (but not always) associated with thrombophlebitis:•pain in the part of the body affected•skin redness or inflammation (not always present)•swelling (edema) of the extremities (ankle and foot)Causes
	Thrombophlebitis is related to a thrombus in the vein.
Infection	An infection is the colonization of a host organism by parasite species. Infecting parasites seek to use the host's resources to reproduce, often resulting in disease. Colloquially, infections are usually considered to be caused by microscopic organisms or microparasites like viruses, prions, bacteria, and viroids, though larger organisms like macroparasites and fungi can also infect.
Pulmonary embolism	Pulmonary embolism is a blockage of the main artery of the lung or one of its branches by a substance that has travelled from elsewhere in the body through the bloodstream (embolism). Usually this is due to embolism of a thrombus (blood clot) from the deep veins in the legs, a process termed venous thromboembolism. A small proportion is due to the embolization of air, fat, talc in drugs of intravenous drug abusers or amniotic fluid.
Breastfeeding	Breastfeeding is the feeding of an infant or young child with breast milk directly from female human breasts (i.e., via lactation) rather than from a baby bottle or other container. Babies have a sucking reflex that enables them to suck and swallow milk. It is recommended that mothers breastfeed for six months or more, without the addition of infant formula or solid food.
Postnatal	Postnatal is the period beginning immediately after the birth of a child and extending for about six weeks. Another term would be postpartum period, as it refers to the mother (whereas postnatal refers to the infant). Less frequently used is puerperium.

Chapter 14. The puerperium

CHAPTER QUIZ: KEY TERMS, PEOPLE, PLACES, CONCEPTS

1. The _____ is a fibromuscular tubular tract which is a sex organ and has two main functions; sexual intercourse and childbirth. In humans, this passage leads from the opening of the vulva to the uterus (womb), but the vaginal tract ends at the cervix. Unlike men, who have only one genital orifice, women have two, the urethra and the _____.

   a. Zona pellucida
   b. Zygote
   c. Portacaval anastomosis
   d. Vagina

2. _____ is a condition where all or part of the placenta or membranes are left behind in the uterus during the third stage of labour.

   In humans, _____ is generally defined as a placenta that has not undergone placental expulsion within 30 minutes of the baby's birth.

   Risks of _____ include hemorrhage and infection.

   a. Rho(D) immune globulin
   b. Retained placenta
   c. Sacrococcygeal teratoma
   d. Justine Siegemund

3. The _____ is the lower, narrow portion of the uterus where it joins with the top end of the vagina. It is cylindrical or conical in shape and protrudes through the upper anterior vaginal wall. Approximately half its length is visible with appropriate medical equipment; the remainder lies above the vagina beyond view.

   a. Cervix
   b. Clitoral body
   c. Clitoral hood
   d. Corona radiata

4. A _____ is a blood clot in a deep vein. A clot inside a blood vessel is called a thrombosis. DVTs predominantly occur in the legs and may have no symptoms.

   a. Deep vein thrombosis
   b. Lymphangitis
   c. Portacaval anastomosis
   d. Thrombophlebitis

5. . _____, is an intricate process in which the skin repairs itself after injury. In normal skin, the epidermis (outermost layer) and dermis (inner or deeper layer) exists in a steady-state equilibrium, forming a protective barrier against the external environment. Once the protective barrier is broken, the normal (physiologic) process of _____ is immediately set in motion.

Chapter 14. The puerperium

CHAPTER QUIZ: KEY TERMS, PEOPLE, PLACES, CONCEPTS

a. magnetic resonance imaging
b. Tissue digestion
c. Wound healing
d. Transversalis fascia

ANSWER KEY
Chapter 14. The puerperium

1. d
2. b
3. a
4. a
5. c

You can take the complete Chapter Practice Test

for Chapter 14. The puerperium
on all key terms, persons, places, and concepts.

Online 99 Cents

http://www.epub5042.32.20612.14.cram101.com/

Use www.Cram101.com for all your study needs

including Cram101's online interactive problem solving labs in

chemistry, statistics, mathematics, and more.

Chapter 15. The transition to neonatal life

CHAPTER OUTLINE: KEY TERMS, PEOPLE, PLACES, CONCEPTS

_____ | Hypoxia

_____ | Vitamin K

_____ | Ductus arteriosus

_____ | Ductus venosus

_____ | Foramen ovale

_____ | Respiratory system

_____ | Surfactant

_____ | Lung

_____ | Fetus

_____ | Adipose tissue

_____ | White adipose tissue

_____ | Hypothermia

_____ | Bilirubin

_____ | Thyroxine

_____ | Triiodothyronine

_____ | Liver

_____ | Metabolism

_____ | Neonatal jaundice

_____ | Meconium

Chapter 15. The transition to neonatal life
CHAPTER OUTLINE: KEY TERMS, PEOPLE, PLACES, CONCEPTS

	Enzyme
	Hormone
	Kidney
	Nervous system
	Immune system
	Vernix caseosa
	Interaction
	Cyanosis
	Hypertension
	Pulmonary hypertension
	Heart development
	Moro reflex

CHAPTER HIGHLIGHTS & NOTES: KEY TERMS, PEOPLE, PLACES, CONCEPTS

Hypoxia	Hypoxia, is a pathological condition in which the body as a whole (generalized hypoxia) or a region of the body (tissue hypoxia) is deprived of adequate oxygen supply. Variations in arterial oxygen concentrations can be part of the normal physiology, for example, during strenuous physical exercise. A mismatch between oxygen supply and its demand at the cellular level may result in a hypoxic condition.

Chapter 15. The transition to neonatal life

CHAPTER HIGHLIGHTS & NOTES: KEY TERMS, PEOPLE, PLACES, CONCEPTS

Vitamin K	Vitamin K is a group of structurally similar, fat-soluble vitamins that are needed for the posttranslational modification of certain proteins required for blood coagulation and in metabolic pathways in bone and other tissue. They are 2-methyl-1,4-naphthoquinone (3-) derivatives. This group of vitamins includes two natural vitamers: vitamin K_1 and vitamin K_2.
Ductus arteriosus	In the developing fetus, the ductus arteriosus also called the ductus Botalli, is a blood vessel connecting the pulmonary artery to the aortic arch. It allows most of the blood from the right ventricle to bypass the fetus's fluid-filled non-functioning lungs. Upon closure at birth, it becomes the ligamentum arteriosum.
Ductus venosus	In the fetus, the ductus venosus shunts approximately half of the blood flow of the umbilical vein directly to the inferior vena cava. Thus, it allows oxygenated blood from the placenta to bypass the liver. In conjunction with the other fetal shunts, the foramen ovale and ductus arteriosus, it plays a critical role in preferentially shunting oxygenated blood to the fetal brain.
Foramen ovale	In the fetal heart, the foramen ovale ( ), also ostium secundum of Born or falx septi, allows blood to enter the left atrium from the right atrium. It is one of two fetal cardiac shunts, the other being the ductus arteriosus (which allows blood that still escapes to the right ventricle to bypass the pulmonary circulation). Another similar adaptation in the fetus is the ductus venosus.
Respiratory system	The respiratory system is the biological system of an organism that introduces respiratory gases to the interior and performs gas exchange. In humans and other mammals, the anatomical features of the respiratory system include airways, lungs, and the respiratory muscles. Molecules of oxygen and carbon dioxide are passively exchanged, by diffusion, between the gaseous external environment and the blood.
Surfactant	Surfactants are compounds that lower the surface tension of a liquid, the interfacial tension between two liquids, or that between a liquid and a solid. Surfactants may act as detergents, wetting agents, emulsifiers, foaming agents, and dispersants. The term surfactant is a blend of surface active agents.
Lung	The lung is the essential respiration cell in many air-breathing animals, including most tetrapods, a few fish and a few snails. In mammals and the more complex life forms, the two lungs are located near the backbone on either side of the heart. Their principal function is to transport oxygen from the atmosphere into the bloodstream, and to release carbon dioxide from the bloodstream into the atmosphere. This exchange of gases is accomplished in the mosaic of specialized cells that form millions of tiny, exceptionally thin-walled air sacs called alveoli.
Fetus	A fetus (sometimes spelled foetus or fœtus) is a stage in the development of viviparous organisms. This stage lies between the embryonic stage and birth.

Chapter 15. The transition to neonatal life

CHAPTER HIGHLIGHTS & NOTES: KEY TERMS, PEOPLE, PLACES, CONCEPTS

Adipose tissue	In histology, adipose tissue is loose connective tissue composed of adipocytes. It is technically composed of roughly only 80% fat; fat in its solitary state exists in the liver and muscles. Adipose tissue is derived from lipoblasts.
White adipose tissue	White adipose tissue or white fat is one of the two types of adipose tissue found in mammals. The other kind of adipose tissue is brown adipose tissue. In humans, white adipose tissue composes as much as 20% of the body weight in men and 25% of the body weight in women.
Hypothermia	Hypothermia is a condition in which core temperature drops below the required temperature for normal metabolism and body functions which is defined as 35.0 °C (95.0 °F). Body temperature is usually maintained near a constant level of 36.5-37.5 °C (98-100 °F) through biologic homeostasis or thermoregulation. If exposed to cold and the internal mechanisms are unable to replenish the heat that is being lost, a drop in core temperature occurs.
Bilirubin	Bilirubin is the yellow breakdown product of normal heme catabolism. Heme is found in hemoglobin, a principal component of red blood cells. Bilirubin is excreted in bile and urine, and elevated levels may indicate certain diseases.
Thyroxine	Thyroxine, or 3,5,3',5'-tetraiodothyronine, a form of thyroid hormones is the major hormone secreted by the follicular cells of the thyroid gland. Synthesis and regulation Thyroxine is synthesized via the iodination and covalent bonding of the phenyl portions of tyrosine residues found in an initial peptide, thyroglobulin, which is secreted into thyroid granules. These iodinated diphenyl compounds are cleaved from their peptide backbone upon being stimulated by thyroid-stimulating hormone.
Triiodothyronine	Triiodothyronine, $C_{15}H_{12}I_3NO_4$, also known as T_3, is a thyroid hormone. It affects almost every physiological process in the body, including growth and development, metabolism, body temperature, and heart rate.
Liver	The liver is a vital organ present in vertebrates and some other animals. It has a wide range of functions, including detoxification, protein synthesis, and production of biochemicals necessary for digestion. The liver is necessary for survival; there is currently no way to compensate for the absence of liver function long term, although liver dialysis can be used short term.
Metabolism	Metabolism is the set of chemical reactions that happen in the cells of living organisms to sustain life.

Chapter 15. The transition to neonatal life

CHAPTER HIGHLIGHTS & NOTES: KEY TERMS, PEOPLE, PLACES, CONCEPTS

	These processes allow organisms to grow and reproduce, maintain their structures, and respond to their environments. The word metabolism can also refer to all chemical reactions that occur in living organisms, including digestion and the transport of substances into and between different cells, in which case the set of reactions within the cells is called intermediary metabolism or intermediate metabolism.
Neonatal jaundice	Neonatal jaundice is a yellowing of the skin and other tissues of a newborn infant. A bilirubin level of more than 85 umol/l (5 mg/dL) manifests clinical jaundice in neonates whereas in adults a level of 34 umol/l (2 mg/dL) would look icteric. In newborns jaundice is detected by blanching the skin with digital pressure so that it reveals underlying skin and subcutaneous tissue.
Meconium	Meconium is the earliest stools of an infant. Unlike later feces, meconium is composed of materials ingested during the time the infant spends in the uterus: intestinal epithelial cells, lanugo, mucus, amniotic fluid, bile, and water. Meconium is almost sterile, unlike later feces, is viscous and sticky like tar, and has no odor.
Enzyme	Enzymes ( ) are biological molecules that catalyze (i.e., increase the rates of) chemical reactions. In enzymatic reactions, the molecules at the beginning of the process, called substrates, are converted into different molecules, called products. Almost all chemical reactions in a biological cell need enzymes in order to occur at rates sufficient for life.
Hormone	A hormone is a chemical released by a cell or a gland in one part of the body that sends out messages that affect cells in other parts of the organism. Only a small amount of hormone is required to alter cell metabolism. In essence, it is a chemical messenger that transports a signal from one cell to another.
Kidney	The kidneys are organs with several functions. They are seen in many types of animals, including vertebrates and some invertebrates. They are an essential part of the urinary system and also serve homeostatic functions such as the regulation of electrolytes, maintenance of acid-base balance, and regulation of blood pressure. They serve the body as a natural filter of the blood, and remove wastes which are diverted to the urinary bladder. In producing urine, the kidneys excrete wastes such as urea and ammonium; the kidneys also are responsible for the reabsorption of water, glucose, and amino acids. The kidneys also produce hormones including calcitriol, renin, and erythropoietin.
Nervous system	The nervous system is an organ system containing a network of specialized cells called neurons that coordinate the actions of an animal and transmit signals between different parts of its body. In most animals the nervous system consists of two parts, central and peripheral. The central nervous system of vertebrates (such as humans) contains the brain, spinal cord, and retina. The peripheral nervous system consists of sensory neurons, clusters of neurons called ganglia, and nerves connecting them to each other and to the central nervous system.

Chapter 15. The transition to neonatal life

CHAPTER HIGHLIGHTS & NOTES: KEY TERMS, PEOPLE, PLACES, CONCEPTS

	These regions are all interconnected by means of complex neural pathways. The enteric nervous system, a subsystem of the peripheral nervous system, has the capacity, even when severed from the rest of the nervous system through its primary connection by the vagus nerve, to function independently in controlling the gastrointestinal system.
Immune system	The immune system is a system of biological structures and processes within an organism that protects against disease. To function properly, an immune system must detect a wide variety of agents, from viruses to parasitic worms, and distinguish them from the organism's own healthy tissue. Pathogens can rapidly evolve and adapt to avoid detection and neutralization by the immune system.
Vernix caseosa	Vernix caseosa, is the waxy or cheese-like white substance found coating the skin of newborn human babies. Vernix has a highly variable makeup but is primarily composed of sebum, cells that have sloughed off the fetus's skin and shed lanugo hair. 12% of the dry weight of vernix is branched-chain fatty acid-containing lipids, cholesterol and ceramide. Vernix of term infants has more squalene and a higher wax ester to sterol ester ratio than preterm infants.
Interaction	Interaction is a kind of action that occurs as two or more objects have an effect upon one another. The idea of a two-way effect is essential in the concept of interaction, as opposed to a one-way causal effect. A closely related term is interconnectivity, which deals with the interactions of interactions within systems: combinations of many simple interactions can lead to surprising emergent phenomena.
Cyanosis	Cyanosis is the appearance of a blue or purple coloration of the skin or mucous membranes due to the tissues near the skin surface being low on oxygen. The onset of cyanosis is 2.5 g/dL of deoxygenated hemoglobin. The bluish color is more readily apparent in those with high hemoglobin counts than it is with those with anemia. Also the bluer color is more difficult to detect on deeply pigmented skin. When signs of cyanosis first appear, such as on the lips or fingers, intervention should be made within 3-5 minutes because a severe hypoxia or severe circulatory failure has induced the cyanosis.
Hypertension	Hypertension or high blood pressure, sometimes called arterial hypertension, is a chronic medical condition in which the blood pressure in the arteries is elevated. This requires the heart to work harder than normal to circulate blood through the blood vessels. Blood pressure involves two measurements, systolic and diastolic, which depend on whether the heart muscle is contracting (systole) or relaxed between beats (diastole).

Chapter 15. The transition to neonatal life

CHAPTER HIGHLIGHTS & NOTES: KEY TERMS, PEOPLE, PLACES, CONCEPTS

Pulmonary hypertension	In medicine, pulmonary hypertension is an increase in blood pressure in the pulmonary artery, pulmonary vein, or pulmonary capillaries, together known as the lung vasculature, leading to shortness of breath, dizziness, fainting, and other symptoms, all of which are exacerbated by exertion. Pulmonary hypertension can be a severe disease with a markedly decreased exercise tolerance and heart failure. It was first identified by Dr. Ernst von Romberg in 1891. According to the most recent classification, it can be one of five different types: arterial, venous, hypoxic, thromboembolic or miscellaneous.
Heart development	The heart is the first functional organ in a vertebrate embryo. There are 5 stages to heart development. The lateral plate mesoderm delaminates to form two layers: the dorsal somatic (parietal) mesoderm and the ventral splanchnic (visceral) mesoderm.
Moro reflex	The Moro reflex, which is distinct from the startle reflex, is one of the infantile reflexes. It may be observed in incomplete form in premature birth after the 28th week of gestation, and is usually present in complete form by week 34 (third trimester). It is normally present in all infants/newborns up to 4 or 5 months of age, and its absence indicates a profound disorder of the motor system.

CHAPTER QUIZ: KEY TERMS, PEOPLE, PLACES, CONCEPTS

1. _____ is the earliest stools of an infant. Unlike later feces, _____ is composed of materials ingested during the time the infant spends in the uterus: intestinal epithelial cells, lanugo, mucus, amniotic fluid, bile, and water. _____ is almost sterile, unlike later feces, is viscous and sticky like tar, and has no odor.

   a. Melena
   b. More Crap
   c. Meconium
   d. Night soil

2. A _____(sometimes spelled foetus or fœtus) is a stage in the development of viviparous organisms. This stage lies between the embryonic stage and birth.

   The fetuses of most mammals are situated similarly to the homo sapiens fetus within their mothers.

   a. Fetus
   b. Gross reproduction rate
   c. Habitual abortion
   d. Human fertilization

Chapter 15. The transition to neonatal life

CHAPTER QUIZ: KEY TERMS, PEOPLE, PLACES, CONCEPTS

3. The _____ is a vital organ present in vertebrates and some other animals. It has a wide range of functions, including detoxification, protein synthesis, and production of biochemicals necessary for digestion. The _____ is necessary for survival; there is currently no way to compensate for the absence of _____ function long term, although _____ dialysis can be used short term.

    a. Diaphragmatic surface of liver
    b. Liver span
    c. Mercapturic acid
    d. Liver

4. _____, is a pathological condition in which the body as a whole (generalized _____) or a region of the body (tissue _____) is deprived of adequate oxygen supply. Variations in arterial oxygen concentrations can be part of the normal physiology, for example, during strenuous physical exercise. A mismatch between oxygen supply and its demand at the cellular level may result in a hypoxic condition.

    a. Mammalian diving reflex
    b. National Board of Diving and Hyperbaric Medical Technology
    c. Hypoxia
    d. Normocapnia

5. _____ or high blood pressure, sometimes called arterial _____, is a chronic medical condition in which the blood pressure in the arteries is elevated. This requires the heart to work harder than normal to circulate blood through the blood vessels. Blood pressure involves two measurements, systolic and diastolic, which depend on whether the heart muscle is contracting (systole) or relaxed between beats (diastole).

    a. Hypertriglyceridemia
    b. Metabolic syndrome
    c. MOMO syndrome
    d. Hypertension

ANSWER KEY
Chapter 15. The transition to neonatal life

1. c
2. a
3. d
4. c
5. d

You can take the complete Chapter Practice Test

for Chapter 15. The transition to neonatal life
on all key terms, persons, places, and concepts.

Online 99 Cents

http://www.epub5042.32.20612.15.cram101.com/

Use www.Cram101.com for all your study needs

including Cram101's online interactive problem solving labs in

chemistry, statistics, mathematics, and more.

Chapter 16. Lactation and infant nutrition

CHAPTER OUTLINE: KEY TERMS, PEOPLE, PLACES, CONCEPTS

_____ Human breast milk

_____ Areola

_____ Nipple

_____ Hypothalamus

_____ Prolactin

_____ Physiology

_____ Domperidone

_____ Dopamine

_____ Biosynthesis

_____ Oxytocin

_____ Lactation

_____ Breast disease

_____ Disease

_____ Menstrual cycle

_____ Ovulation

_____ Fertility

_____ Maternal nutrition

_____ Obesity

_____ Colostrum

Chapter 16. Lactation and infant nutrition
CHAPTER OUTLINE: KEY TERMS, PEOPLE, PLACES, CONCEPTS

_____ Large intestine

_____ Amino acid

_____ Essential amino acid

_____ Non-protein nitrogen

_____ Fatty acid

_____ Arachidonic acid

_____ Cholesterol

_____ Amylase

_____ Galactose

_____ Oligosaccharide

_____ Vitamin D

_____ Vitamin E

_____ Gut-associated lymphoid tissue

_____ Vital capacity

_____ Vitamin K

_____ Fibronectin

_____ Sex steroid

_____ Blood cell

_____ Premenstrual syndrome

Chapter 16. Lactation and infant nutrition

CHAPTER OUTLINE: KEY TERMS, PEOPLE, PLACES, CONCEPTS

- _____ Infertility
- _____ Hormone replacement therapy
- _____ Endocrine system
- _____ Sickle-cell disease
- _____ Dominance
- _____ Fragile X syndrome
- _____ Growth factor
- _____ Pre-eclampsia
- _____ Chorionic villus sampling
- _____ Osteoclast
- _____ Ejaculatory duct
- _____ Fetus
- _____ Immune system
- _____ Marfan syndrome
- _____ Blood pressure
- _____ Monitoring
- _____ Diaphragm
- _____ Body mass index
- _____ Salmonellosis

Chapter 16. Lactation and infant nutrition
CHAPTER OUTLINE: KEY TERMS, PEOPLE, PLACES, CONCEPTS

	Presentation
	Asphyxia
	Lochia
	Soft tissue
	Infection
	Postnatal

CHAPTER HIGHLIGHTS & NOTES: KEY TERMS, PEOPLE, PLACES, CONCEPTS

Human breast milk	Breast milk, to be specific human milk, is the milk produced by the breasts (or mammary glands) of a human female for her infant offspring. Milk is the primary source of nutrition for newborns before they are able to eat and digest other foods; older infants and toddlers may continue to be breastfed, either exclusively or in combination with other foods.
	The baby nursing from its own mother is the most ordinary way of obtaining breastmilk, but the milk can be pumped and then fed by baby bottle, cup and/or spoon, supplementation drip system, and nasogastric tube. Breastmilk can be supplied by a woman other than the baby's mother; either via donated pumped milk (for example from a milk bank), or when a woman nurses a child other than her own at her breast -- an ancient and storied practice known as wetnursing.
	The World Health Organization recommends exclusive breastfeeding for the first six months of life, with solids gradually being introduced around this age when signs of readiness are shown. Supplemented breastfeeding is recommended until at least age two and then for as long as the mother and child wish.
	Breastfeeding continues to offer health benefits into and after toddlerhood.

Chapter 16. Lactation and infant nutrition

CHAPTER HIGHLIGHTS & NOTES: KEY TERMS, PEOPLE, PLACES, CONCEPTS

These benefits include a somewhat lowered risk of Sudden Infant Death Syndrome (SIDS), increased intelligence, decreased likelihood of contracting middle ear infections, cold, and flu bugs, a tiny decrease in the risk of childhood leukemia, lower risk of childhood onset diabetes, decreased risk of asthma and eczema, decreased dental problems and decreased risk of obesity later in life, and may possibly include a decreased risk of developing psychological disorders, particularly in adopted children.

Breastfeeding also provides health benefits for the mother. It assists the uterus in returning to its pre-pregnancy size and reduces post-partum bleeding, as well as assisting the mother in returning to her pre-pregnancy weight. Breastfeeding also reduces the risk of breast cancer later in life. Production

Under the influence of the hormones prolactin and oxytocin, women produce milk after childbirth to feed the baby. The initial milk produced is often referred to as colostrum, which is high in the immunoglobulin IgA, which coats the gastrointestinal tract. This helps to protect the newborn until its own immune system is functioning properly, and creates a mild laxative effect, expelling meconium and helping to prevent the build-up of bilirubin (a contributory factor in jaundice).

Actual inability to produce enough milk is rare, with studies showing that mothers from developing countries experiencing nutritional hardship still produce amounts of milk of similar quality to that of mothers in developed countries. There are many reasons a mother may not produce enough breast milk. Some of the most common are an improper latch (i.e., the baby does not connect efficiently with the nipple), not nursing or pumping enough to meet supply, certain medications (including estrogen-containing hormonal contraceptives), illness, and dehydration. A rarer reason is Sheehan's syndrome, also known as postpartum hypopituitarism, which is associated with prolactin deficiency; this syndrome may require hormone replacement.

The amount of milk produced depends on how often the mother is nursing and/or pumping; the more the mother nurses her baby, or pumps, the more milk is produced. It is very helpful to nurse on demand - to nurse when the baby wants to nurse rather than on a schedule. If pumping, it is helpful to have an electric high-grade pump so that all of the milk ducts are stimulated. Some mothers try to increase their milk supply in other ways - by taking the herb fenugreek, used for hundreds of years to increase supply ; there are also prescription medications that can be used, such as Domperidone (off-label use) and Reglan. Composition

The exact integrated properties of breast milk are not entirely understood, but the nutrient content after this period is relatively consistent and draws its ingredients from the mother's food supply.

Areola	In anatomy, an areola, plural areolæ is any circular area on the breast such as the colored skin surrounding the nipple. Although the term is most commonly used to describe the pigmented area around the human nipple (areola mammae), it can also be used to describe other small circular areas such as the inflamed region surrounding a pimple.

Chapter 16. Lactation and infant nutrition

CHAPTER HIGHLIGHTS & NOTES: KEY TERMS, PEOPLE, PLACES, CONCEPTS

Nipple	In its most general form, a nipple is a structure from which a fluid emanates. More specifically, it is the projection on the breasts or udder of a mammal by which breast milk is delivered to a mother's young. In this sense, it is often called a teat, especially when referring to non-humans, and the medical term used to refer to it is papilla.
Hypothalamus	The hypothalamus is a portion of the brain that contains a number of small nuclei with a variety of functions. One of the most important functions of the hypothalamus is to link the nervous system to the endocrine system via the pituitary gland (hypophysis). The hypothalamus is located below the thalamus, just above the brain stem.
Prolactin	Prolactin also known as luteotropic hormone (LTH) is a protein that in humans is encoded by the PRL gene. Prolactin is a peptide hormone discovered by Henry Friesen. Although it is perhaps best known for its role in lactation, prolactin already existed in the oldest known vertebrates--fish--where its most important functions were probably related to control of water and salt balance.
Physiology	Physiology is the science of the function of living systems. This includes how organisms, organ systems, organs, cells, and bio-molecules carry out the chemical or physical functions that exist in a living system. The highest honor awarded in physiology is the Nobel Prize in Physiology or Medicine, awarded since 1901 by the Royal Swedish Academy of Sciences.
Domperidone	Domperidone is an antidopaminergic drug, developed by Janssen Pharmaceutica, and used orally, rectally or intravenously, generally to suppress nausea and vomiting, as a prokinetic agent and for promoting lactation. Gastrointestinal problems There is some evidence that domperidone has antiemetic activity. Domperidone is used, together with metoclopramide, cyclizine, and $5HT_3$ receptor antagonists (such as granisetron) in the treatment of nausea and vomiting.
Dopamine	Dopamine is a simple organic chemical in the catecholamine family, is a monoamine neurotransmitter which plays a number of important physiological roles in the bodies of animals. In addition to being a catecholamine and a monoamine, dopamine may be classified as a substituted phenethylamine. Its name derives from its chemical structure, which consists of an amine group (NH_2) linked to a catechol structure called dihydroxyphenethylamine, the decarboxylated form of dihydroxyphenylalanine (acronym DOPA).
Biosynthesis	Biosynthesis is an enzyme-catalyzed process in cells of living organisms by which substrates are converted to more complex products. The biosynthesis process often consists of several enzymatic steps in which the product of one step is used as substrate in the following step.

Chapter 16. Lactation and infant nutrition

CHAPTER HIGHLIGHTS & NOTES: KEY TERMS, PEOPLE, PLACES, CONCEPTS

Oxytocin	Oxytocin ( ) is a mammalian hormone that acts primarily as a neuromodulator in the brain.
	Oxytocin is best known for its roles in sexual reproduction, in particular during and after childbirth. It is released in large amounts after distension of the cervix and uterus during labor, facilitating birth, and after stimulation of the nipples, facilitating breastfeeding.
Lactation	Lactation describes the secretion of milk from the mammary glands and the period of time that a mother lactates to feed her young. The process occurs in all female mammals, however it predates mammals. In humans the process of feeding milk is called breastfeeding or nursing.
Breast disease	Breast diseases can be classified either with disorders of the integuement, or disorders of the reproductive system. A majority of breast diseases are noncancerous.
	Breast awareness is a goal of the breast health movement.
Disease	A disease is an abnormal condition affecting the body of an organism. It is often construed to be a medical condition associated with specific symptoms and signs. It may be caused by external factors, such as infectious disease, or it may be caused by internal dysfunctions, such as autoimmune diseases.
Menstrual cycle	The menstrual cycle is the scientific term for the physiological changes that can occur in fertile women for the purposes of sexual reproduction and fertilization
	The menstrual cycle, under the control of the endocrine system, is necessary for reproduction.
Ovulation	Ovulation is the process in a female's menstrual cycle by which a mature ovarian follicle ruptures and discharges an ovum (also known as an oocyte, female gamete, or casually, an egg). Ovulation also occurs in the estrous cycle of other female mammals, which differs in many fundamental ways from the menstrual cycle. The time immediately surrounding ovulation is referred to as the ovulatory phase or the periovulatory period.
Fertility	Fertility is the natural capability of producing offsprings. As a measure, 'fertility rate' is the number of children born per couple, person or population. Fertility differs from fecundity, which is defined as the potential for reproduction (influenced by gamete production, fertilisation and carrying a pregnancy to term).
Maternal nutrition	Maternal nutrition is the dietary intake and habits of expectant mothers with dual emphasis on the health of the mother and the physical and mental development of infants. Nearly 24 per cent of babies are estimated to be born with lower than optimal weights at birth.

Chapter 16. Lactation and infant nutrition

CHAPTER HIGHLIGHTS & NOTES: KEY TERMS, PEOPLE, PLACES, CONCEPTS

Obesity	Obesity is a medical condition in which excess body fat has accumulated to the extent that it may have an adverse effect on health, leading to reduced life expectancy and/or increased health problems. Body mass index (BMI), a measurement which compares weight and height, defines people as overweight (pre-obese) if their BMI is between 25 and 30 kg/m^2, and obese when it is greater than 30 kg/m^2. Obesity increases the likelihood of various diseases, particularly heart disease, type 2 diabetes, obstructive sleep apnea, certain types of cancer, and osteoarthritis.
Colostrum	Colostrum is a form of milk produced by the mammary glands of mammals in late pregnancy. Most species will generate colostrum just prior to giving birth. Colostrum contains antibodies to protect the newborn against disease, as well as being lower in fat and higher in protein than ordinary milk.
Large intestine	The large intestine is the third-to-last part of the digestive system in vertebrate animals.
Amino acid	Amino acids are molecules containing an amine group, a carboxylic acid group and a side-chain that varies between different amino acids. The key elements of an amino acid are carbon, hydrogen, oxygen, and nitrogen. They are particularly important in biochemistry, where the term usually refers to alpha-amino acids.
Essential amino acid	An essential amino acid is an amino acid that cannot be synthesized de novo by the organism (usually referring to humans), and therefore must be supplied in the diet. (*) Essential only in certain cases. (**) Pyrrolysine, sometimes considered 'the 22nd amino acid', is not listed here as it is not used by humans.
Non-protein nitrogen	Non-protein nitrogen is a term used in animal nutrition to refer collectively to components such as urea, biuret, and ammonia, which are not proteins but can be converted into proteins by microbes in the ruminant stomach. Due to their lower cost compared to plant and animal proteins their inclusion in a diet can result in economic gain, but at too high levels cause a depression in growth and possible ammonia toxicity (microbes convert NPN to ammonia first before using that to make protein). NPN can also be used to artificially raise crude protein values, which are measured based on nitrogen content, as protein is about 16% nitrogen, but, for example, urea is 47% nitrogen.
Fatty acid	In chemistry, especially biochemistry, a fatty acid is a carboxylic acid with a long aliphatic tail (chain), which is either saturated or unsaturated. Most naturally occurring fatty acids have a chain of an even number of carbon atoms, from 4 to 28.

Chapter 16. Lactation and infant nutrition

CHAPTER HIGHLIGHTS & NOTES: KEY TERMS, PEOPLE, PLACES, CONCEPTS

	Fatty acids are usually derived from triglycerides or phospholipids. When they are not attached to other molecules, they are known as 'free' fatty acids.
Arachidonic acid	Arachidonic acid is a polyunsaturated omega-6 fatty acid 20:4(ω-6). It is the counterpart to the saturated arachidic acid found in peanut oil, (L. arachis - peanut).
	Arachidonic acid is one of the essential fatty acids required by most mammals. Some mammals lack the ability to--or have a very limited capacity to--convert linoleic acid into arachidonic acid, making it an essential part of their diet. Since little or no arachidonic acid is found in common plants, such animals are obligate carnivores; the cat is a common example. A commercial source of arachidonic acid has been derived, however, from the fungus Mortierella alpina
Cholesterol	Cholesterol is an organic chemical substance classified as a waxy steroid of fat. It is an essential structural component of mammalian cell membranes and is required to establish proper membrane permeability and fluidity.
	In addition to its importance within cells, cholesterol is an important component in the hormonal systems of the body for the manufacture of bile acids, steroid hormones, and vitamin D. Cholesterol is the principal sterol synthesized by animals; in vertebrates it is formed predominantly in the liver.
Amylase	Amylase is an enzyme that catalyses the breakdown of starch into sugars. Amylase is present in human saliva, where it begins the chemical process of digestion. Food that contains much starch but little sugar, such as rice and potato, taste slightly sweet as they are chewed because amylase turns some of their starch into sugar in the mouth.
Galactose	Galactose, is a type of sugar that is less sweet than glucose. It is a C-4 epimer of glucose.
	Galactan is a polymer of the sugar galactose found in hemicellulose.
Oligosaccharide	An oligosaccharide is a saccharide polymer containing a small number (typically two to ten) of component sugars, also known as simple sugars (monosaccharides). Oligosaccharides can have many functions; for example, they are commonly found on the plasma membrane of animal cells where they can play a role in cell-cell recognition.
	In general, they are found either O- or N-linked to compatible amino acid side-chains in proteins or to lipid moieties .
Vitamin D	Vitamin D is a group of fat-soluble secosteroids, the two major physiologically relevant forms of which are vitamin D_2 (ergocalciferol) and vitamin D_3 (cholecalciferol). Vitamin D without a subscript refers to either D_2 or D_3 or both.

Chapter 16. Lactation and infant nutrition

CHAPTER HIGHLIGHTS & NOTES: KEY TERMS, PEOPLE, PLACES, CONCEPTS

Vitamin E	Vitamin E is used to refer to a group of fat-soluble compounds that include both tocopherols and tocotrienols. There are many different forms of vitamin E, of which γ-tocopherol is the most abundant in the North American diet. γ-Tocopherol can be found in corn oil, soybean oil, margarine and dressings.
Gut-associated lymphoid tissue	The digestive tract's immune system is often referred to as gut-associated lymphoid tissue and works to protect the body from invasion. GALT is an example of mucosa-associated lymphoid tissue.
	The digestive tract is an important component of the body's immune system.
Vital capacity	Vital capacity is the maximum amount of air a person can expel from the lungs after a maximum inhalation. It is equal to the inspiratory reserve volume plus the tidal volume plus the expiratory reserve volume.
	A person's vital capacity can be measured by a spirometer which can be a wet or regular spirometer.
Vitamin K	Vitamin K is a group of structurally similar, fat-soluble vitamins that are needed for the posttranslational modification of certain proteins required for blood coagulation and in metabolic pathways in bone and other tissue. They are 2-methyl-1,4-naphthoquinone (3-) derivatives. This group of vitamins includes two natural vitamers: vitamin K_1 and vitamin K_2.
Fibronectin	Fibronectin is a high-molecular weight (~440kDa) extracellular matrix glycoprotein that binds to membrane-spanning receptor proteins called integrins. In addition to integrins, fibronectin also binds extracellular matrix components such as collagen, fibrin and heparan sulfate proteoglycans (e.g. syndecans).
	Fibronectin exists as a dimer, consisting of two nearly identical monomers linked by a pair of disulfide bonds. The fibronectin protein is produced from a single gene, but alternative splicing of its pre-mRNA leads to the creation of several isoforms.
Sex steroid	Sex steroids, also known as gonadal steroids, are steroid hormones that interact with vertebrate androgen or estrogen receptors. Their effects are mediated by slow genomic mechanisms through nuclear receptors as well as by fast nongenomic mechanisms through membrane-associated receptors and signaling cascades. The term sex hormone is nearly always synonymous with sex steroid.
Blood cell	A blood cell, is a cell produced by haematopoiesis and normally found in blood. In mammals, these fall into three general categories:•Red blood cells -- Erythrocytes•White blood cells -- Leukocytes•Platelets -- Thrombocytes.

Chapter 16. Lactation and infant nutrition

CHAPTER HIGHLIGHTS & NOTES: KEY TERMS, PEOPLE, PLACES, CONCEPTS

Together, these three kinds of blood cells add up to a total 45% of the blood tissue by volume, with the remaining 55% of the volume composed of plasma, the liquid component of blood. This volume percentage (e.g., 45%) of cells to total volume is called hematocrit, determined by centrifuge or flow cytometry.

Premenstrual syndrome
Premenstrual syndrome (also called PMT or premenstrual tension) is a collection of physical and emotional symptoms related to a woman's menstrual cycle. While most women of child-bearing age (up to 85%) report having experienced physical symptoms related to normal ovulatory function, such as bloating or breast tenderness, medical definitions of PMS are limited to a consistent pattern of emotional and physical symptoms occurring only during the luteal phase of the menstrual cycle that are of 'sufficient severity to interfere with some aspects of life'. In particular, emotional symptoms must be present consistently to diagnose PMS. The specific emotional and physical symptoms attributable to PMS vary from woman to woman, but each individual woman's pattern of symptoms is predictable, occurs consistently during the ten days prior to menses, and vanishes either shortly before or shortly after the start of menstrual flow.

Infertility
Infertility primarily refers to the biological inability of a person to contribute to conception. Infertility may also refer to the state of a woman who is unable to carry a pregnancy to full term. There are many biological causes of infertility, some which may be bypassed with medical intervention.

Hormone replacement therapy
Hormone replacement therapy (HRT) is a system of medical treatment for surgically menopausal, transgender, perimenopausal and postmenopausal women. It is based on the idea that the treatment may prevent discomfort caused by diminished circulating oestrogen and progesterone hormones, and in the case of the surgically or prematurely menopausal, that it may prolong life and may reduce incidence of dementia. It involves the use of one or more of a group of medications designed to artificially boost hormone levels.

Endocrine system
The endocrine system is the system of glands, each of which secretes a type of hormone directly into the bloodstream to regulate the body. The endocrine system is in contrast to the exocrine system, which secretes its chemicals using ducts. It derives from the Greek words 'endo' meaning inside, within, and 'crinis' for secrete.

Sickle-cell disease
Sickle-cell disease is an autosomal recessive genetic blood disorder, with overdominance, characterized by red blood cells that assume an abnormal, rigid, sickle shape. Sickling decreases the cells' flexibility and results in a risk of various complications. The sickling occurs because of a mutation in the haemoglobin gene. Life expectancy is shortened, with studies reporting an average life expectancy of 42 in males and 48 in females.

Dominance
Dominance in genetics is a relationship between two variant forms (alleles) of a single gene, in which one allele masks the expression of the other in influencing some trait.

Chapter 16. Lactation and infant nutrition

CHAPTER HIGHLIGHTS & NOTES: KEY TERMS, PEOPLE, PLACES, CONCEPTS

	In the simplest case, if a gene exists in two allelic forms (A & B), three combinations of alleles (genotypes) are possible: AA, AB, and BB. If AB individuals (heterozygotes) show the same form of the trait (phenotype) as AA individuals (homozygotes), and BB homozygotes show an alternative phenotype, allele A is said to dominate or be dominant to allele B, and B is said to be recessive to A.
Fragile X syndrome	Fragile X syndrome Martin-Bell syndrome, or Escalante's syndrome (more commonly used in South American countries), is a genetic syndrome that is the most common known single-gene cause of autism and the most common inherited cause of mental retardation among boys. It results in a spectrum of intellectual disability ranging from mild to severe as well as physical characteristics such as an elongated face, large or protruding ears, and larger testes (macroorchidism), behavioral characteristics such as stereotypic movements (e.g. hand-flapping), and social anxiety.
	Fragile X syndrome is associated with the expansion of the CGG trinucleotide repeat affecting the Fragile X mental retardation 1 (FMR1) gene on the X chromosome, resulting in a failure to express the fragile X mental retardation protein (FMRP), which is required for normal neural development.
Growth factor	A growth factor is a naturally occurring substance capable of stimulating cellular growth, proliferation and cellular differentiation. Usually it is a protein or a steroid hormone. Growth factors are important for regulating a variety of cellular processes.
Pre-eclampsia	Pre-eclampsia is a medical condition in which hypertension arises in pregnancy (pregnancy-induced hypertension) in association with significant amounts of protein in the urine.
	Pre-eclampsia refers to a set of symptoms rather than any causative factor, and there are many different causes for the condition. It appears likely that there are substances from the placenta that can cause endothelial dysfunction in the maternal blood vessels of susceptible women.
Chorionic villus sampling	Chorionic villus sampling sometimes misspelled 'chorionic villous sampling', is a form of prenatal diagnosis to determine chromosomal or genetic disorders in the fetus. It entails sampling of the chorionic villus (placental tissue) and testing it for chromosomal abnormalities, usually with FISH or PCR. chorionic\ villus\ sampling usually takes place at 10-12 weeks' gestation, earlier than amniocentesis (14-16 weeks). It is the preferred technique before 15 weeks.
Osteoclast	An osteoclast (from the Greek words for 'bone' and 'broken') is a type of bone cell that removes bone tissue by removing its mineralized matrix and breaking up the organic bone (organic dry weight is 90% collagen). This process is known as bone resorption. Osteoclasts were discovered by Kolliker in 1873.

Chapter 16. Lactation and infant nutrition

CHAPTER HIGHLIGHTS & NOTES: KEY TERMS, PEOPLE, PLACES, CONCEPTS

Ejaculatory duct	The ejaculatory ducts (ductus ejaculatorii) are paired structures in male anatomy. Each ejaculatory duct is formed by the union of the vas deferens with the duct of the seminal vesicle. They pass through the prostate, and open into the urethra at the Colliculus seminalis.
Fetus	A fetus (sometimes spelled foetus or fœtus) is a stage in the development of viviparous organisms. This stage lies between the embryonic stage and birth. The fetuses of most mammals are situated similarly to the homo sapiens fetus within their mothers.
Immune system	The immune system is a system of biological structures and processes within an organism that protects against disease. To function properly, an immune system must detect a wide variety of agents, from viruses to parasitic worms, and distinguish them from the organism's own healthy tissue. Pathogens can rapidly evolve and adapt to avoid detection and neutralization by the immune system.
Marfan syndrome	Marfan syndrome is a genetic disorder of the connective tissue. People with Marfan's tend to be unusually tall, with long limbs and long, thin fingers. It is inherited as a dominant trait.
Blood pressure	Blood pressure is the pressure exerted by circulating blood upon the walls of blood vessels, and is one of the principal vital signs. When used without further specification, 'blood pressure' usually refers to the arterial pressure of the systemic circulation. During each heartbeat, blood pressure varies between a maximum (systolic) and a minimum (diastolic) pressure.
Monitoring	In medicine, monitoring is the evaluation of a disease or condition over time. It can be performed by continuously measuring certain parameters (for example, by continuously measuring vital signs by a bedside monitor), and/or by repeatedly performing medical tests (such as blood glucose monitoring in people with diabetes mellitus). Transmitting data from a monitor to a distant monitoring station is known as telemetry or biotelemetry.
Diaphragm	The diaphragm is a cervical barrier type of birth control. It is a soft latex or silicone dome with a spring molded into the rim. The spring creates a seal against the walls of the vagina.

Chapter 16. Lactation and infant nutrition

CHAPTER HIGHLIGHTS & NOTES: KEY TERMS, PEOPLE, PLACES, CONCEPTS

Body mass index	The body mass index or Quetelet index, is a heuristic proxy for human body fat based on an individual's weight and height. BMI does not actually measure the percentage of body fat. It was devised between 1830 and 1850 by the Belgian polymath Adolphe Quetelet during the course of developing 'social physics'.
Salmonellosis	Salmonellosis is an infection with Salmonella bacteria. Most people infected with Salmonella develop diarrhea, fever, vomiting, and abdominal cramps 12 to 72 hours after infection. In most cases, the illness lasts four to seven days, and most people recover without treatment.
Presentation	In obstetrics, the presentation of a fetus about to be born refers to which anatomical part of the fetus is leading, that is, is closest to the pelvic inlet of the birth canal. According to the leading part, this is identified as a cephalic, breech, or shoulder presentation. A malpresentation is any other presentation than a vertex presentation.
Asphyxia	Asphyxia is a condition of severely deficient supply of oxygen to the body that arises from being unable to breathe normally. An example of asphyxia is choking. Asphyxia causes generalized hypoxia, which primarily affects the tissues and organs.
Lochia	In the field of obstetrics, lochia is post-partum vaginal discharge, containing blood, mucus, and placental tissue. Lochia discharge typically continues for 4 to 6 weeks after childbirth. It progresses through three stages.
Soft tissue	In anatomy, the term soft tissue refers to tissues that connect, support, or surround other structures and organs of the body, not being bone. Soft tissue includes tendons, ligaments, fascia, skin, fibrous tissues, fat, and synovial membranes (which are connective tissue), and muscles, nerves and blood vessels (which are not connective tissue). It is sometimes defined by what it is not.
Infection	An infection is the colonization of a host organism by parasite species. Infecting parasites seek to use the host's resources to reproduce, often resulting in disease. Colloquially, infections are usually considered to be caused by microscopic organisms or microparasites like viruses, prions, bacteria, and viroids, though larger organisms like macroparasites and fungi can also infect.
Postnatal	Postnatal is the period beginning immediately after the birth of a child and extending for about six weeks. Another term would be postpartum period, as it refers to the mother (whereas postnatal refers to the infant). Less frequently used is puerperium.

Chapter 16. Lactation and infant nutrition

CHAPTER QUIZ: KEY TERMS, PEOPLE, PLACES, CONCEPTS

1. _____ is a genetic disorder of the connective tissue. People with Marfan's tend to be unusually tall, with long limbs and long, thin fingers.

   It is inherited as a dominant trait.

   a. magnetic resonance imaging
   b. Marfan syndrome
   c. Immunization
   d. Immunocompetence

2. _____s, also known as gonadal steroids, are steroid hormones that interact with vertebrate androgen or estrogen receptors. Their effects are mediated by slow genomic mechanisms through nuclear receptors as well as by fast nongenomic mechanisms through membrane-associated receptors and signaling cascades. The term sex hormone is nearly always synonymous with _____.

   a. Transition nuclear protein
   b. Sex steroid
   c. Vagina
   d. Zona pellucida

3. _____ is an enzyme that catalyses the breakdown of starch into sugars. _____ is present in human saliva, where it begins the chemical process of digestion. Food that contains much starch but little sugar, such as rice and potato, taste slightly sweet as they are chewed because _____ turns some of their starch into sugar in the mouth.

   a. Anticaking agent
   b. Aversive agent
   c. Azodicarbonamide
   d. Amylase

4. An _____ is an amino acid that cannot be synthesized de novo by the organism (usually referring to humans), and therefore must be supplied in the diet.

   (*) Essential only in certain cases.

   (**) Pyrrolysine, sometimes considered 'the 22nd amino acid', is not listed here as it is not used by humans.

   a. Essential nutrient
   b. Essential amino acid
   c. European Prospective Investigation into Cancer and Nutrition
   d. European Society for Clinical Nutrition and Metabolism

5. . _____ (also called PMT or premenstrual tension) is a collection of physical and emotional symptoms related to a woman's menstrual cycle.

Chapter 16. Lactation and infant nutrition

CHAPTER QUIZ: KEY TERMS, PEOPLE, PLACES, CONCEPTS

While most women of child-bearing age (up to 85%) report having experienced physical symptoms related to normal ovulatory function, such as bloating or breast tenderness, medical definitions of PMS are limited to a consistent pattern of emotional and physical symptoms occurring only during the luteal phase of the menstrual cycle that are of 'sufficient severity to interfere with some aspects of life'. In particular, emotional symptoms must be present consistently to diagnose PMS. The specific emotional and physical symptoms attributable to PMS vary from woman to woman, but each individual woman's pattern of symptoms is predictable, occurs consistently during the ten days prior to menses, and vanishes either shortly before or shortly after the start of menstrual flow.

a. Ritushuddhi
b. Sanitary napkin
c. Premenstrual syndrome
d. The Story of Menstruation

ANSWER KEY
Chapter 16. Lactation and infant nutrition

1. b
2. b
3. d
4. b
5. c

You can take the complete Chapter Practice Test

for Chapter 16. Lactation and infant nutrition
on all key terms, persons, places, and concepts.

Online 99 Cents

http://www.epub5042.32.20612.16.cram101.com/

Use www.Cram101.com for all your study needs

including Cram101's online interactive problem solving labs in

chemistry, statistics, mathematics, and more.

Other Cram101 e-Books and Tests

Want More?
Cram101.com...

Cram101.com provides the outlines and highlights of your textbooks, just like this e-StudyGuide, but also gives you the PRACTICE TESTS, and other exclusive study tools for all of your textbooks.

Learn More. *Just click*
http://www.cram101.com/

Other Cram101 e-Books and Tests